# METALLURGY - ADVANCES IN MATERIALS AND PROCESSES

# METALLURGY - ADVANCES IN MATERIALS AND PROCESSES

*Editor*

**Teodor Piatek**

scitus
academics

**Metallurgy - Advances in Materials and Processes**
Edited by **Teodor Piatek**

ISBN: 978-1-68117-641-3
Library of Congress Control Number: 2016946502



**Reprint 2025**

SCITUS Academics LLC,
616, Corporate Way, Suite 2, 4766,
Valley Cottage, NY 10989
www.scitusacademics.com



**Notice**



Printed in United States of America on Acid Free Paper

# PREFACE

Metallurgy is a domain of materials science and of materials engineering that studies the physical and chemical behavior of metallic elements and their mixtures, which are called alloys. Modern metallurgy stems from the ancient desire to understand fully the behavior of metals. Long ago, the art of the metalworker was enshrouded in mystery and folklore. In modern times scientists and engineers around the world have been responding to the requisite of high performance materials over ground-breaking material research and engineering. The ever growing demand on quality and reliability has caused in some dazzling technological accomplishments in the arena of advanced materials and manufacturing. Nearly everything we need for our present-day civilization depends on metals. Vast quantities of steels, aluminum, titanium, copper, and nickel alloys are used for automobiles, ships, aircraft, spacecraft, bridges, and buildings as well as the machines required to produce them. Almost all uses of electricity depend on copper and aluminum. All around us we see the utilization of aluminum, copper, and steels, often in new applications combining metals with plastics and fiber-reinforced composite materials. Some metals such as titanium and zirconium – impossible to smelt or extract from ores just a few years ago – are now used in large quantities and referred to as space-age metals. Metallurgy is relevant to the materials science, welding, machine shop, quality control, and industrial technology industries, each of which share equal responsibility for the design, development and implementation of metals and materials processing in industry today.

*Metallurgy - Advances in Materials and Processes* is to bring together significant findings of leading experts, in developing and improving the technology that supports advanced materials and process development. It is envisioned that it will stimulate knowledge transfer across the materials society together with university students, engineers and scientists to build further understanding of the subject.

# TABLE OF CONTENTS


Chapter 1. Heat Treatment of Dental Alloys: A Review 1

Chapter 2. Effect of Processing Parameters on the Mechanical Properties of Heavy Section Ductile Iron 27

Chapter 3. Microstructure-Property Relationship in Advanced Ni-Based Superalloys 47

Chapter 4. Gold Nanostructures Prepared on Solid Surface 77

Chapter 5. Characterization of Precipitates in a Microalloyed Steel Using Quantitative X-ray Diffraction 113

Chapter 6. Artificial Neural Networks Investigation of Indentation Force Effects on Nano- and Microhardness of Dual Phase Steels 137

Chapter 7. Aluminium AA6061 Matrix Composite Reinforced with Spherical Alumina Particles Produced by Infiltration: Perspective on Aerospace Applications 157

Chapter 8. Microgalvanic Corrosion Behavior of Cu-Ag Active Braze Alloys Investigated with SKPFM 179

Chapter 9. Homogenization Heat Treatment to Reduce the Failure of Heat Resistant Steel Castings 207

Chapter 10. Development and Characterization of a Metal Injection Molding Bio Sourced Inconel 718 Feedstock Based on Polyhydroxyalkanoates 229

Chapter 11. Mould Filling Ability Characterisation of Cast Aluminium Alloys Using Design of Experiments 245

Chapter 12. Preparation of Metallic Iron Powder from Pyrite Cinder by Carbothermic Reduction and Magnetic Separation 271

Index 289

1

# HEAT TREATMENT OF DENTAL ALLOYS: A REVIEW

**William A. Brantley[1] and Satish B. Alapati[2]**

[1] Division of Restorative, Prosthetic and Primary Care Dentistry, Graduate Program in Dental Materials Science, College of Dentistry, The Ohio State University, Columbus, OH, USA

[2] Department of Endodontics, College of Dentistry, University of Illinois at Chicago, Chicago, IL, USA

## 1. INTRODUCTION

Metallic materials have widespread use in dentistry for clinical treatment and restoration of teeth. Major areas of usage are: (1) restorative dentistry and prosthodontics (dental amalgam and gold alloy restorations for single teeth, metallic restorations for multiple teeth, including metal-ceramic restorations, removable partial denture frameworks, and dental implants), (2) orthodontics (wires which provide the biomechanical force for tooth movement), and (3) endodontics (rotary and hand instruments for treatment of root canals). Heat treatment of the metal can be performed by the manufacturer, dental laboratory, or dentist to alter properties intentionally and improve clinical performance. Heat treatment of the metal also occurs during the normal sequence of preparing a metal-ceramic restoration, when dental porcelain is bonded to the underlying alloy substrate. Moreover, intraoral heat treatment of some metallic restorations occurs over long periods of time. There is an enormous scientific literature on the heat treatment of metals for dentistry. A search of the biomedical literature in May 2012, using PubMed [http://www.ncbi. nlm. nih.gov/pubmed/] revealed nearly 450 articles on heat treatment of dental alloys. The purpose of this chapter is to provide a review of the heat treatment of metallic dental materials in the foregoing important areas, describing the important property changes, with a focus on the underlying metallurgical principles.

# 2. RESTORATIVE DENTISTRY AND PROSTHODONTICS

## 2.1. Dental Amalgams

Dental amalgams are prepared in the dental office by mixing particles of a silver-tin-copper alloy for dental amalgam that may contain other trace metals with liquid mercury. The initially mixed (termed triturated) material is in a moldable condition and is placed (termed condensed) directly by the dentist into the prepared tooth cavity, where it undergoes a setting process that produces multiple phases and can require up to one day for near completion. Extensive information about the several different types of dental amalgams are provided in textbooks on dental materials [1,2]. Particles of the alloy for dental amalgam are manufactured by either lathe-cutting a cast ingot or directing the molten alloy through a special nozzle. Both the machining of the lathe-cut particles and the rapid solidification of the spherical particles create residual stress. In addition, the microstructure of the solidified silver-tin-copper alloy has substantial microsegregation. Consequently, manufacturers of the alloy powder for dental amalgam perform a proprietary heat treatment to relieve residual stresses and obtain a more homogeneous microstructure. This heat treatment is of considerable practical importance since it affects the setting time of the dental amalgam after the powder is mixed with mercury. Subsequently, the dental amalgam restorations undergo intraoral aging, which can be regarded as heat treatment, and detailed information about the microstructural phase changes for prolonged intraoral time periods has been obtained from clinically retrieved dental amalgam restorations [3].

## 2.2. Gold Alloys For All-Metal Restorations

Gold alloys are principally used for all-metal restorations (inlays, crowns and onlays) in single posterior teeth. These alloys are cast by a precision investment process, and the restorations are cemented by the dentist into the prepared tooth cavity. The original gold casting alloys contained over approximately 70 wt.% gold, but the very high price of gold has led to the development of alloys that contain approximately 50 wt.% gold. These alloys also contain silver, copper, platinum, palladium, zinc, and other trace elements, including iridium for grain refinement. Information about the dental casting process and the gold alloys is available in dental materials textbooks [1,2]. Detailed compositions and mechanical properties of specific alloys are available on the website of the major manufacturers. Another valuable reference is the current ISO Standard on metallic materials for fixed and removable dental appliances [4], which stipulates mechanical property requirements. In the normal dental laboratory procedure, gold castings for all-metal restorations are water-quenched after solidification, following loss of the red heat appearance for the sprue. This results in formation of a disordered substitutional solid solution and leaves the alloy in the soft condition, which is preferable since adjustments are more easily made on the restoration by the dental laboratory or dentist. The gold alloy casting can also be placed in the soft condition by heating at 700°C for 15 minutes and water-quenching. The quenched gold casting may be placed in the hard condition by heat treatment at 350°C for 15 minutes and air-cooling. This heat treatment results in formation of ordered AuCu or $AuCu_3$ regions in the disordered matrix of the high-gold or lower-gold alloys, respectively. Examples of changes in clinically important mechanical properties from heat treatment are shown in Table 1 for two gold alloys, where (S) and (H) represent the soft and hard conditions.

In practice, dental laboratories do not perform heat treatments on the cast gold restorations because of the time involved. However, it appears to be fortunate that the gold alloys that contain sufficient copper to undergo ordering will undergo age hardening in the mouth. Figure 1 compares the intraoral aging behavior of a traditional high-gold dental alloy (Type IV) and a special gold alloy containing gallium (AuCu-3wt%Ga) [5].

## 2.3. Alloys For Fixed Prosthodontics (Metal-Ceramic Restorations)

Metal-ceramic restorations are in widespread clinical use for restorative and prosthetic dentistry, and are employed for single-tooth restorations and for restorations involving multiple adjacent teeth (fixed prostheses or crown-and-bridgework). An alloy is cast using the precision investment procedure in dental laboratories to fit accurately to the prepared tooth or teeth, and to form a substrate (termed the coping) for the porcelain. After an initial oxidation step that forms a native oxide on the metal surface, one or two layers of opaque porcelain are bonded to the metal, followed by the application of a layer of body porcelain and a surface glaze [1,2]. In order to have a strong bond between the porcelain and metal, which is essential for clinical longevity of the metal-ceramic restoration, the coefficients of thermal contraction for the metal and porcelain must be closely matched, and a difference not exceeding 0.5 ppm/°C is generally desired. Mechanical property requirements for the alloys are stipulated in ANSI/ADA Specification No. 38 (ISO 9693) [6], and the minimum value of 250 MPa for the 0.2% offset yield strength is important, since the thin coping must withstand intraoral forces without undergoing permanent deformation. The metal-ceramic bond strength (termed the bond compatibility index) is measured with a three-point bending test that uses thin cast alloy strip specimens having a centrally located area of sintered porcelain, and a minimum bond strength (shear stress) of 25 MPa is stipulated.

**Table 1.** Summary of property changes resulting from heat treatment of two gold alloys for all-metal restorations. [http://www.jelenko.com/, accessed August 15, 2012]

| Alloy | Vickers Hardness | | 0.2% Offset Yield Strength | | Percentage Elongation | |
|---|---|---|---|---|---|---|
| Firmilay (74.5% Au) | 121 (S) | 182 (H) | 207 MPa (S) | 276 MPa (H) | 39% (S) | 19% (H) |
| Midas (46% Au) | 135 (S) | 230 (H) | 345 MPa (S) | 579 MPa (H) | 30% (S) | 13% (H) |

Both noble and base metal alloys are used for bonding to dental porcelain. The current American Dental Association classification has four alloy groups for fixed prosthodontics [7]: (1) high-noble (gold-platinum-palladium, gold-palladium-silver and gold-palladium); (2) noble (palladium-silver, palladium-copper-gallium, and palladium-gallium); (3) predominantly base metal (nickel-chromium and cobalt-chromium); (4) titanium and titanium alloys. Information about these alloys for metal-ceramic bonding is summarized in a textbook on fixed prosthodontics [8]. The principal mechanisms for metal-ceramic bonding are (a) mechanical interlocking from the initially viscous porcelain at the elevated sintering temperatures flowing into microirregularities on the air-abraded cast metal surface and (b) chemical bonding associated with an interfacial oxide layer between the metal and ceramic. These two mechanisms are evident from photomicrographs, found in numerous references [8], of the fracture surfaces for metal-ceramic specimens prepared from a wide variety of dental alloys. This native oxide forms on the cast alloy during the initial oxidation firing step in the dental porcelain furnace. Noble alloys for bonding to dental porcelain contain small amounts of secondary elements, such as tin, indium and iron, which form the native oxide and also increase the alloy strength. However, Mackert et al [9] found that during initial oxidation heat treatment, metallic Pd-Ag nodules formed on the surface of a palladium-silver alloy for metal-ceramic restorations and only internal oxidation occurred for the tin and indium present in the alloy composition. They concluded that porcelain bonding arose predominantly from mechanical interlocking with the nodules. Internal oxidation has also been reported for high-gold [10] and high-palladium [11] alloys for bonding to porcelain, but both alloy types also formed surface oxides [10,12].

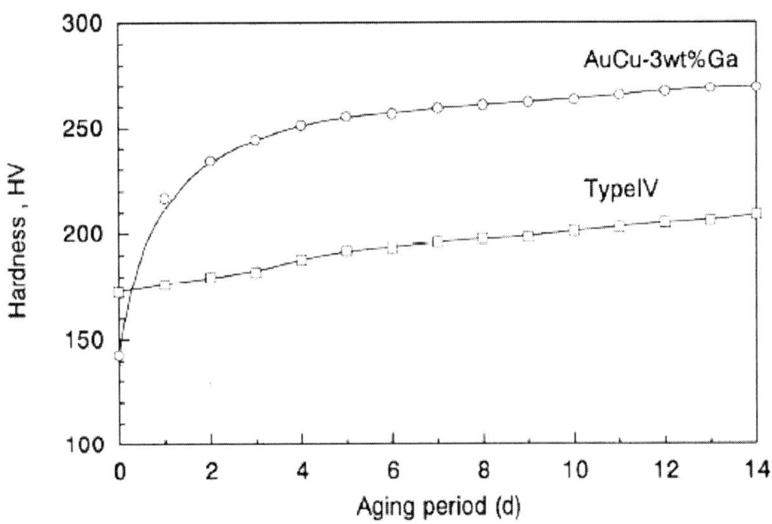


**Figure 1.** Comparison of the two-week aging behavior at 37°C for a high-gold dental alloy and a dental gold alloy containing gallium that was designed to undergo intraoral aging. From [5] and reproduced with permission.

The initial oxidation step and subsequent sintering (also termed baking or firing) of the dental porcelain layers causes the alloy to experience substantial heat-treatment effects. Under normal dental laboratory conditions, the porcelain firing sequence is performed rapidly. For example, in one study heating of high-palladium alloys in the dental porcelain furnace was performed at approximately 30°C/min over a temperature range from 650°C to above 900°C, and the total heating time for the several firing cycles at these elevated temperatures was about 45 minutes [11]. Studies [13-15] have shown that the as-cast microstructures of noble metal alloys for bonding to porcelain are highly inhomogeneous in the initial as-cast condition, presumably from substantial elemental microsegregation that occurs during the rapid solidification involved with casting into much cooler investment [1,2]. After simulation of the dental porcelain firing sequence, the noble metal alloy microstructures become substantially homogeneous, and there are accompanying changes in the mechanical properties, as shown in Table 2.

Peaks in Vickers hardness for heat treatments at temperatures that span the porcelain-firing temperature range indicate that influential precipitation processes can occur in some noble alloys for fixed prosthodontics [13,16]. For the gold-palladium-silver alloy in Table 1, heating an as-cast specimen to 980°C caused a pronounced decrease in Vickers hardness, and subsequent heat treatments at temperatures from 200° to 980°C revealed a pronounced peak in Vickers hardness at approximately 760°C. The absence of substantial changes in Vickers hardness for similar heat treatments of the gold-palladium alloy in Table 2 arises from differences in the precipitates that form in the two complex alloy compositions. Figure 2 presents the age hardening behavior of a palladium-silver alloy, where specimens were subjected to isothermal annealing for 30 minute time periods at temperatures from 400°C to 900°C that span the range for the porcelain firing cycles [16]. Bulk values of Vickers hardness were obtained with 1 kg loads, and 25 g loads were used to obtain hardness values for specific microstructural regions. In contrast, research suggests that microstructures of popular nickel-chromium base metal alloys used with dental porcelain are not changed substantially during dental laboratory processing [17].

## 2.4. Alloys For Removable Prosthodontics

Base metal casting alloys (nickel-chromium, cobalt-chromium and cobalt-chromium-nickel) are popular for fabricating the metallic frameworks for removable partial dentures because of their lower cost [1,2]. Once an active area of dental metallurgy research, studies have found that these alloys have dendritic microstructures in the as-cast condition, because of the absence of suitable grain-refining elements, and that heat treatment is ineffective for producing improved mechanical properties [18]. A more recent publication shows the dendritic microstructures of some current alloys and their mechanical properties [19].

**Table 2.** Mechanical properties for two noble metal alloy types used with dental porcelain, comparing the as-cast condition (C) and simulated porcelain firing heat treatment (F) [13].

| Alloy Type | Vickers Hardness | | 0.2% Offset Yield Strength | | Percentage Elongation | |
|---|---|---|---|---|---|---|
| Au-Pd-Ag (Neydium) | 199 (C) | 218 (P) | 420 MPa (C) | 490 MPa (F) | 6% (C) | 8% (F) |
| Au-Pd (Olympia) | 213 (C) | 225 (P) | 500 MPa (C) | 540 MPa (F) | 13% (C) | 20% (F) |

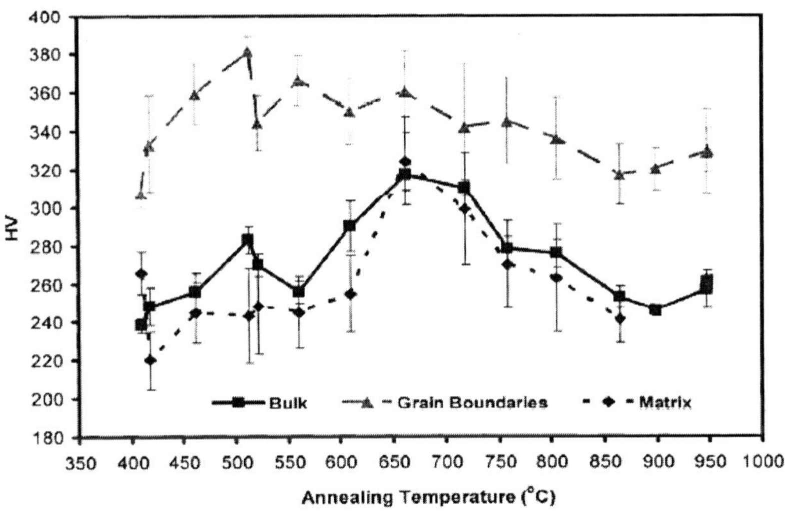


**Figure 2.** Annealing behavior of a palladium-silver alloy for fixed prosthodontics, showing changes in Vickers hardness for a heat treatment temperature range that spans the porcelain firing cycles. Reproduced from [16] with permission.

Removable partial denture frameworks have clasps that engage the teeth. These clasps can be cast as part of the entire framework, or alternatively wire clasps can be joined to the cast framework in the dental laboratory [1,2]. Both noble metal and base metal wires for clasps are available [20]. Because of their superior strength compared to the cast base metal alloys, wire clasps with smaller cross-section dimensions can be used with the frameworks, but caution is required during joining in the dental laboratory to avoid overheating that will cause loss of the wrought microstructure. Wire clasps are used in the as-received condition; heat treatment is not recommended before joining to the framework.

## 2.5. Dental Implant Alloys

Dental implants in current widespread clinical use are manufactured from CP (commercially pure) titanium or Ti-6Al-4V, and some implants have a thin bioceramic surface coating (typically hydroxyapatite, the principal inorganic constituent of bone and tooth structure). Proprietary heat treatments [21] are performed on Ti-6Al-4V by manufacturers to obtain optimum microstructures for the implants; minimal information is currently available about these microstructures in the dental scientific literature.

Recently, there has been considerable research interest in the development of new titanium implant alloys for orthopedic applications that have improved biocompatibility compared to the Ti-6Al-4V alloy in widespread current use. There is particular interest in the beta-titanium alloys which have lower elastic modulus than Ti-6Al-4V to minimize stress shielding and subsequent loss of the surrounding bone which has a much lower elastic modulus. Stress shielding does not seem to be of concern for dental implants, presumably because of the threaded designs. Biocompatible titanium-niobium-zirconium beta alloys have been investigated, and oxide nanotubes can be grown on the alloy surface by an anodization technique, and subsequent heat treatment can be employed to modify the structure of the nanotubes [22]. In another exciting research area, titanium oxide nanowires have been recently grown on both CP titanium and Ti-6Al-4V using special elevated-temperature oxidation heat treatments in an argon atmosphere with low oxygen concentrations [23]. Both of these special types of surface oxide layers may prove to be useful for dental and orthopedic implants, but future testing in animals will be needed to examine their efficacy.

# 3. ORTHODONTICS

## 3.1. Background

Orthodontic wires engaged in brackets that are bonded to teeth, after being deformed elastically during initial placement, provide the biomechanical force for tooth movement during unloading. There are four wire types in current clinical practice: stainless steel, cobalt-chromium, beta-titanium and nickel-titanium [24]. The clinically important mechanical properties are (a) elastic modulus, which is proportional to the biomechanical force when wires of similar dimensions are compared; (b) springback, which is generally expressed as the quotient of yield strength and elastic modulus (YS/E), and represents the approximate strain at the end of the clinically important elastic range; and (c) modulus of resilience, expressed as $YS^2/2E$ and representing the spring energy available for tooth movement. (The permanent deformation portion of orthodontic wire activation is ineffective for tooth movement.) Round orthodontic wires are manufactured by a proprietary drawing sequence that involves several stages with intermediate annealing heat treatments. Rectangular orthodontic wires are manufactured by a rolling process utilizing a Turk's head apparatus. The wire drawing process with the heat treatments greatly affects mechanical properties.

## 3.2. Stainless Steel Orthodontic Wires

A recent study that investigated stainless steel wires used in orthodontic practice found that most products were AISI Type 304 and that AISI Type 316L (low carbon) and nickel-free ASTM Type F2229 were also available [25]. While standard physical metallurgy textbooks consider the elastic modulus to be a structure-insensitive property, research has shown that the permanent deformation and heat treatments involved with the wire drawing process can substantially affect the elastic modulus of stainless steel orthodontic wires [26,27]. X-ray diffraction has revealed that conventional orthodontic wires manufactured from AISI Types 302 and 304, while predominantly austenitic structure, can contain the $\alpha'$ martensitic phase, depending upon the carbon content and temperatures involved with the processing [28]. The presence of this martensitic phase accounts for the reduction in elastic modulus for some conventional stainless steel orthodontic wires. In addition, when fabricating complex stainless steel appliances, it is recommended that orthodontists perform a stress-relief heat treatment to prevent fracture during manipulation; a heating time up to 15 minutes and a temperature range of 300° to 500°C appears to be acceptable [29-31]. Heating austenitic stainless steel to temperatures between 400° and 900°C can result in chromium carbide precipitation at grain boundaries and cause the alloy to become susceptible to intergranular corrosion, and heating of austenitic stainless steel wires above 650°C should not be done because loss of the wrought microstructure causes degradation of mechanical properties.

## 3.3. Cobalt-Chromium Orthodontic Wires

The cobalt-chromium orthodontic wire (Elgiloy) marketed by Rocky Mountain Orthodontics (Denver, CO, USA) contains 40% Co, 20% Cr, 15.81% Fe, 15% Ni, 7% Mo, 2% Mn, 0.15% C carbon and 0.04% Be beryllium (https://www.rmortho.com/, accessed August 15, 2012). Four different tempers (spring quality) are available, and the soft Blue temper is favored by many orthodontists because the wire is easily manipulated in the as-received condition, and then heat treated to increase the yield strength and modulus of resilience. Heat treatment (not recommended for the most resilient temper) is conveniently performed with the electrical resistance welding apparatus commonly used in orthodontic practice, and the manufacturer provides a special paste that indicates when the heat treatment is complete. Alternatively, furnace heat treatment performed at 480°C for 5 minutes has been found to give satisfactory results [32]. An extensive study employing furnace heat treatment (480°C for 10 minutes) for three tempers and numerous sizes of the Elgiloy wires observed increases of 10% – 20% in elastic modulus and 10% – 20% in 0.1% offset yield strength, which resulted in substantial improvement of the modulus of resilience [27]. These changes in mechanical properties arise from complex precipitation processes during heat treatment that are not understood. Many other companies now market cobalt-chromium orthodontic wires, but studies of their mechanical properties and the results of heat treatment have not been reported.

## 3.4. Beta-Titanium and Other Titanium-Based Orthodontic Wires

Beta-titanium orthodontic wires have the advantages of: (a) known biocompatibility from the absence of nickel in the alloy composition; (b) lower elastic modulus than stainless steel and cobalt-chromium wires, which provides more desirable lower orthodontic force for tooth movement; (c) higher springback than stainless steel and cobalt-chromium wires, which is desirable for the archwire to have greater elastic range; and (d) high formability and weldability, which are needed for fabrication of certain appliances [24]. A recent study [25] of commercially available titanium-based orthodontic wires revealed that most products are Beta III alloys [21] containing approximately 11.5 Mo, 6 Zr, and 4.5 Sn, similar to the original beta-titanium wire introduced to orthodontics [33,34]. Beta C [21] and Ti-45Nb beta-titanium and Ti-6Al-4V (alpha-beta) wire products are also available [25]. Heat treatment is not performed by the orthodontist on these wires, but care with the wire drawing and intermediate heat treatments by the manufacturer are essential for obtaining the desired mechanical properties. These processes must be conducted under well-controlled conditions because of the highly reactive nature of titanium.

## 3.5. Nickel-Titanium Orthodontic Wires

Following the pioneering work of Andreasen and his colleagues [35,36], near-equiatomic nickel-titanium (NiTi) wire was introduced to orthodontics by the Unitek Corporation (now 3M Unitek) [37]. This wire had the advantages of a much lower elastic modulus than the stainless steel and cobalt-chromium wires available at the time and a very large elastic range. The clinical disadvantage is that substantial permanent deformation of this wire is not possible to obtain certain orthodontic appliances that can be fabricated with the three preceding, highly formable, alloys. The original nickel-titanium wire had a work-hardened martensitic structure and did not exhibit the superelastic behavior (termed pseudoelasticity in engineering materials science) or the true shape memory characteristics displayed by subsequently introduced NiTi wires [1,38-41]. These nickel-titanium wires have been a very active area of research.

The mechanical properties of the nickel-titanium orthodontic wires are determined by the proportions and character of three microstructural phases: (a) austenite, which occurs under conditions of high temperature and low stress; (b) martensite, which occurs under conditions of low temperature and high stress; and (c) R-phase, which forms as an intermediate phase during the transformation between martensite and austenite. Very careful control of the wire processing and associated heat treatments, along with precise compositional control, by the manufacturer are needed to produce nickel-titanium wires with the desired superelastic, nonsuperelastic, or shape memory character [42,43].

Heat treatments have been exploited by manufacturers to control the orthodontic force ranges produced by nickel-titanium archwires [39]. Heat treatment temperatures have ranged from 400° to 600°C with times from 5 minutes to 2 hours [39,40]. Effects of heat treatment on cantilever bending plots for two sizes of a round superelastic nickel-titanium wire are presented in Figure 3 [40].

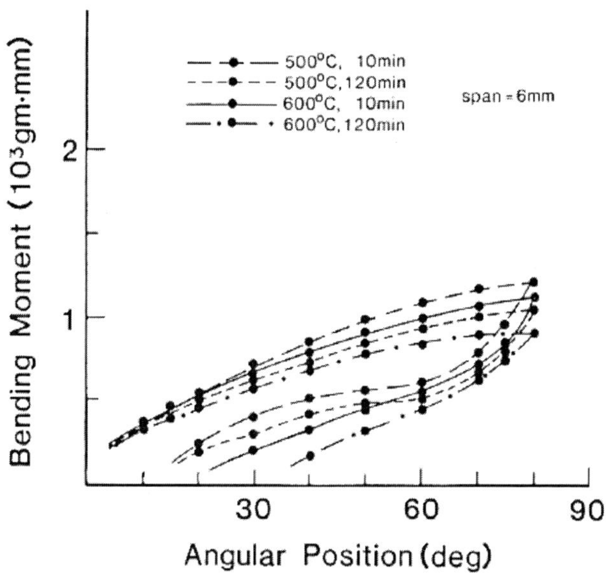


**Figure 3.** Effects of heat treatments on cantilever bending plots for 6 mm test spans of a superelastic nickel-titanium orthodontic wire. Reproduced from [40] with permission.

Loss of superelastic behavior occurs for the 2 hour heat treatment at 600°C, evidenced by the large decrease in springback (difference between the original deflection of 80 degrees and the final angular position on unloading). Heat treatment at 500°C for 10 minutes had minimal effect, while heat treatment for 2 hours caused a decrease in the average superelastic bending moment during the unloading region of clinical importance. Bending properties for nonsuperelastic wires were only slightly affected by these heat treatments. In addition to the use of furnace heat treatment, electrical resistance heat treatment [44] has also been exploited by one manufacturer to produce archwires where the level of biomechanical force varies with position along the wire [24].

Microstructural phases at varying temperatures in nickel-titanium orthodontic wires and their transformations are conveniently studied by differential scanning calorimetry (DSC) [45]. Temperature-modulated DSC provides greater insight into the transformations than conventional DSC [46]. Figures 4and 5 present temperature-modulated DSC heating curves for shape memory and superelastic nickel-titanium orthodontic wires, respectively. The transformations involving austenite (A), martensite (M) and R-phase (R) are labeled. The austenite-finish ($A_f$) temperature for completion of the transformation from martensite to austenite on heating is determined by the intersection with the adjacent baseline of a tangent line to the peak for the final transformation to austenite [47].

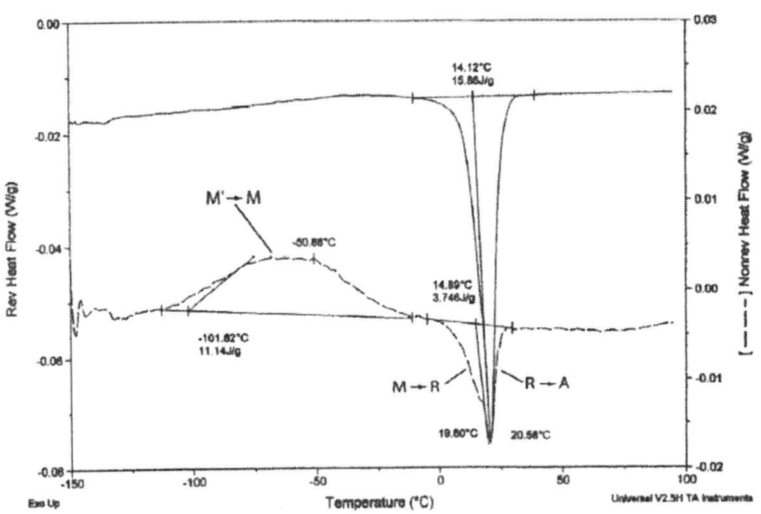


**Figure 4.** Heating temperature-modulated DSC plot for a shape memory nickel-titanium orthodontic wire. Reproduced from [46] with permission.

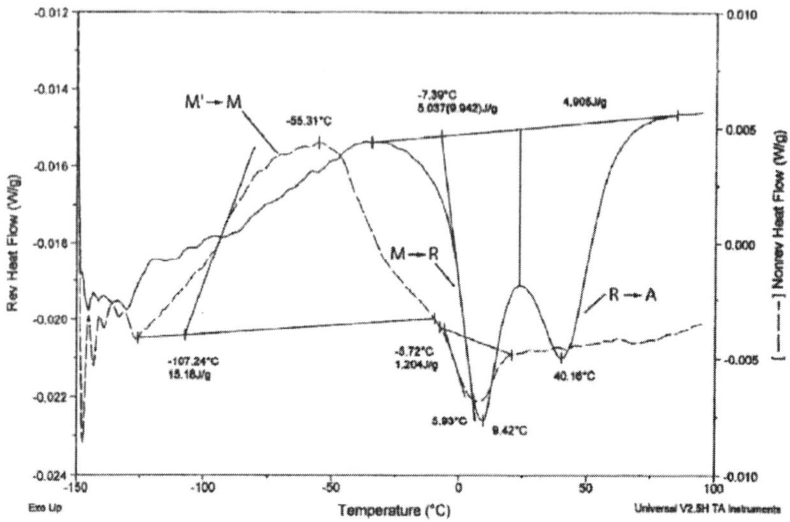


**Figure 5.** Heating temperature-modulated DSC plot for a superelastic nickel-titanium orthodontic wire. Reproduced from [46] with permission.

The $A_f$ temperature is below body temperature (37°C) for nickel-titanium wires that exhibit shape memory in the oral environment. The superelastic nickel-titanium wires have $A_f$ temperatures that are greater than mouth temperature and have more widely separated peaks for the successive transformations from M →R and R → A. The nonsuperelastic wires have much weaker transformations (lower values of enthalpy [ΔH]) and $A_f$ temperatures that are also greater than mouth temperature [45]. Examination of x-ray diffraction patterns for nickel-titanium orthodontic wires revealed the effects of heat treatment on the $M_s$ temperature for the start of the cooling transformation to martensite as well as the occurrence of stress relief and perhaps some recrystallization [24,48].

Transformation of a low temperature martensite phase (M′) to the higher temperature form of martensite (M), shown in Figures 5 and 6, is readily detected as a large exothermic peak on the nonreversing heat flow curves from temperature-modulated DSC. Transmission electron microscopy has revealed that this transformation arise from low-temperature twinning within the martensite structure [49].

# 4. ENDODONTICS

## 4.1. Stainless Steel Instruments

Traditionally, endodontic treatment was performed with stainless steel hand files and reamers to remove the injured or diseased dental pulp from the root canals of teeth. While conventional elevated-temperature heat treatment is not recommended for these instruments, they are subjected to sterilization procedures before being using again with a different patient. One study found that dry heat sterilization (180°C for 2 hours) and autoclave sterilization (220 kPa pressure and 136°C for 10 minutes) slightly decreased the flexibility and resistance to torsional fracture of the instruments but they still satisfied the requirements for minimum angular deflection in the ISO standard [50]. Further research is needed to gain insight into the metallurgical origins of the property changes.

## 4.2. Nickel-Titanium Instruments

Following the pioneering work of Walia et al that introduced the nickel-titanium hand file to the endodontics profession [51], engine-driven rotary instruments were introduced that enable rapid instrumentation of root canals. These instruments are in widespread clinical use, and research on the nickel-titanium files has been a highly intensive area of research.

The major mechanical property of the equiatomic nickel-titanium alloy that led to replacement of the traditional austenitic stainless steel files was the much lower elastic modulus of NiTi, which enabled curved root canals to be negotiated with facility. An excellent review article [52] describes the manufacturing process for the nickel-titanium files, which are generally machined from starting wire blanks. The conventional nickel-titanium rotary instruments have been fabricated from superelastic nickel-titanium blanks.

Defects caused by the machining process and metallurgical flaws in the starting blanks, along with inadvertent overloading by the clinician, can result in fracture of the file within the root canal, which causes considerable patient anguish since the broken fragments often cannot be easily retrieved [53,54].

A recent study investigated the effect of heat treatment on conventional nickel-titanium rotary instruments, using temperature-modulated DSC and Micro-X-ray diffraction [55]. Results are shown inFigure 6 (a) – (d) for heat treatment at temperatures from 400° to 800°C in a flowing nitrogen atmosphere.

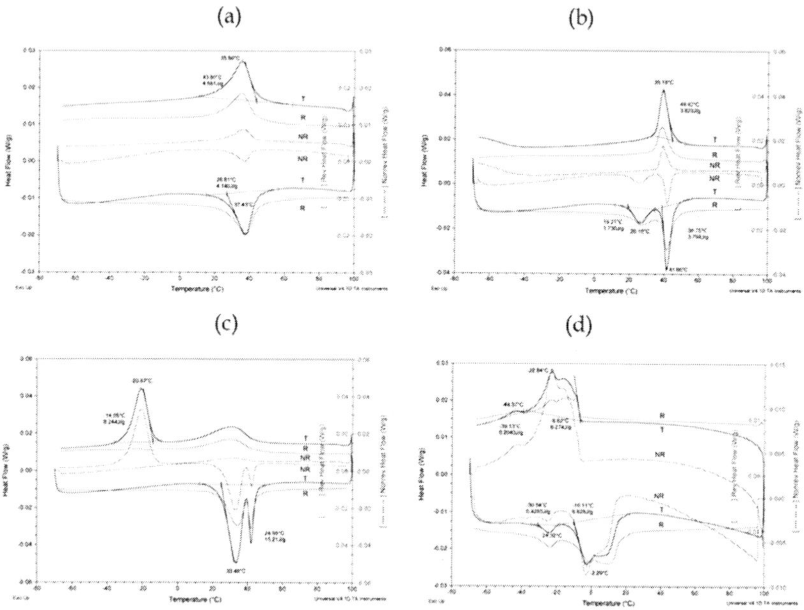


**Figure 6.** Temperature-modulated DSC reversing (R), nonreversing (NR) and total (T) heat flow curves for specimens from conventional rotary endodontic instruments after heat treatment in flowing nitrogen for 15 minutes at (a) 400°, (b) 500°, (c) 600° and (d) 850°C. From [55] and reproduced with permission.

Heat treatment between 400° and 600°C increased the $A_f$ temperature for as-received conventional NiTi rotary instruments to approximately 45° – 50°C, and the transformations between martensite and austenite were changed to a more reversing character than nonreversing character [55]. Heat treatment in a nitrogen atmosphere might lead to a harder surface from the formation of nitrides [56], which is beneficial for cutting efficiency of the rotary instrument. This research suggested that heat treatment at temperatures near 500°C in a nitrogen atmosphere might yield the optimum microstructure and mechanical properties, with improved resistance to deformation and fracture for conventional NiTi rotary instruments. Heat treatment at temperatures exceeding 600°C should not be performed, since the superelastic behavior is lost along with potential degradation of the wrought microstructure [24]. Another study has reported that heat treatment at 430° and 440°C greatly improved the fatigue resistance of one conventional rotary instrument product [57].

New nickel-titanium rotary instruments have been marketed, for which the wire blanks were improved by special proprietary processing techniques, including heat treatment. The first notable example was M-Wire, named for its stable martensitic structure [58]. Previous conventional rotary instruments were fabricated from superelastic wire blanks with evident transformable austenite detected by conventional DSC [59]. However, when the conventional instruments were cooled far below room temperature to attain the fully martensite condition, the enthalpy changes for transformations from martensite to austenite were far below those for superelastic orthodontic wires [44,45], indicating that these instruments contain a substantial proportion of stable martensite in their microstructures.

Two different batches of M-Wire (termed Type 1 and Type 2), with unknown differences in proprietary processing, were obtained for characterization by temperature-modulated DSC and Micro-X-ray diffraction [58]. Figure 7 shows the differences in the temperature-modulated DSC plots for (a) conventional superelastic wire and (b) Type 1 M-Wire.

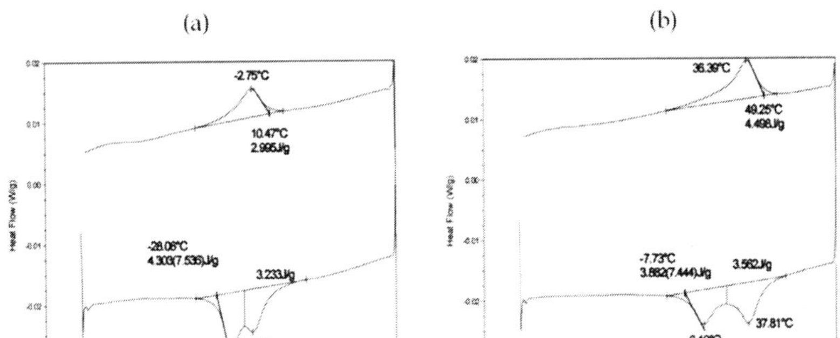


**Figure 7.** Comparison of temperature-modulated DSC total heat flow for (a) conventional superelastic wire and (b) Type 1 M-Wire. Lower curves are the plots for the heating cycles. Reproduced from [58] with permission.

The general appearances of the temperature-modulated DSC plots in Figure 7 (a) and (b) are similar. However, the approximate $A_f$ temperatures for the conventional superelastic wire and Type 1 M-Wire were approximately 15°C and 50°C, respectively. The approximate $A_f$ temperature for the Type 2 M-Wire was 45°C. The proportions of the different NiTi phases were quite different for Type 1 and Type 2 M-Wire, as shown in Figure 8.

The Micro-X-ray diffraction pattern indicated that Type 1 M-Wire had an austenitic structure, and the Micro-X-ray diffraction pattern from the conventional superelastic wire was similar. In contrast, the Micro-X-ray diffraction pattern from Type 2 M-Wire contained additional peaks for martensite and R-phase, along with peaks for austenite. However, when M-Wire was examined by transmission electron microscopy, a heavily deformed martensitic structure was found [58]. The explanation is that the DSC peaks only reveal NiTi phases that are capable of undergoing transformation and that (stable) heavily deformed martensitic NiTi only produces weak x-ray diffraction peaks. Rotary instruments fabricated from M-Wire have been found to have similar $A_f$ values, microstructures and Vickers hardness, so the machining process and other proprietary fabrication steps do not appear to markedly alter the inherent structure and properties of the starting blanks [60].

Recently, new nickel-titanium rotary instruments have been introduced, in which the wire blank is heated to an appropriate temperature for transformation to the R-phase and twisted, along with repeated heat treatment and other subsequent thermal processing; instruments have been characterized by conventional DSC and cantilever bending tests [61]. Another recent study has characterized several new nickel-titanium rotary instruments by DSC and conventional x-ray diffraction, along with optical and scanning electron microscopic examination of their microstructures, including use of energy-dispersive x-ray spectroscopic analyses (SEM/EDS), to investigate the martensitic microstructures and composition of precipitates [62]. Because of the potentially great commercial importance, development of new rotary instruments with improved clinical performance is expected to remain an area of intensive research, along with study of the role of heat treatment [63].

It is essential to appreciate the complexity of the physical metallurgy of the nickel-titanium alloys and the effects of the severe thermomechanical processing of the starting wire blanks, along with heat treatments and machining of the wire blanks, on the metallurgical structure. Transmission electron microscopy and electron diffraction remain the best techniques to gain insight into the instrument microstructures and elucidate the relationships with mechanical properties and clinical performance.

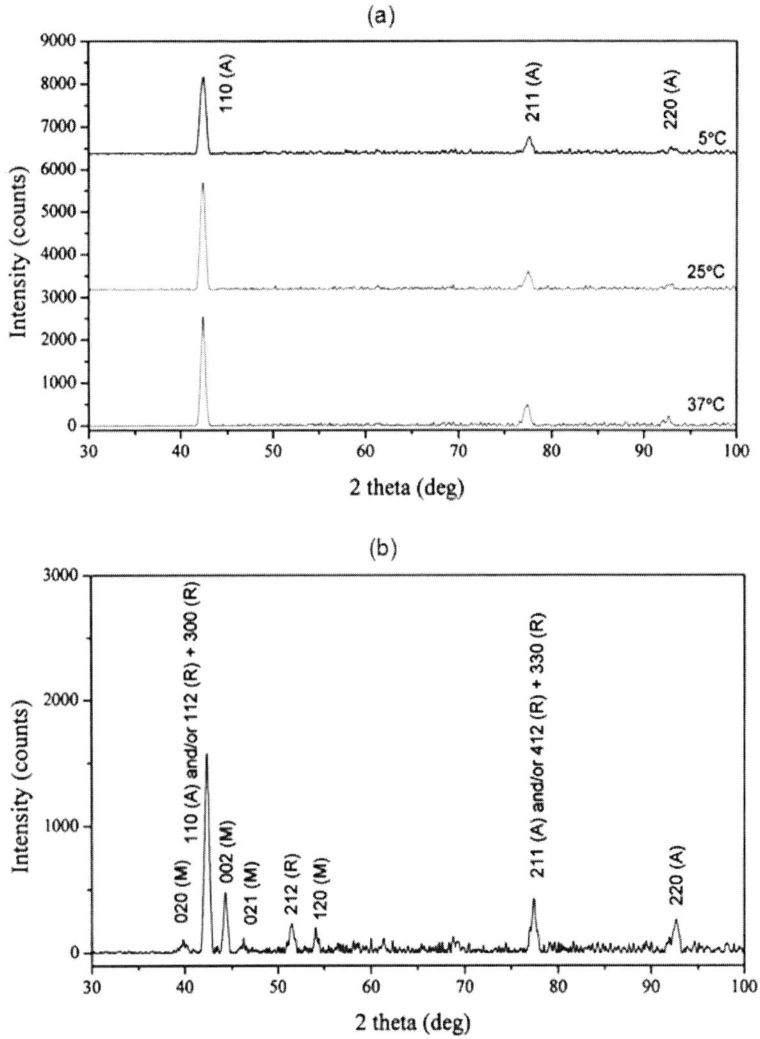


**Figure 8.** Micro-X-ray diffraction patterns for (a) Type 1 M-Wire and (b) Type 2 M-Wire. Peaks for austenite (A), martensite (M) and R-phase (R) are labeled. Reproduced from [58] with permission.

# REFERENCES


1. K. J. Anusavice, (editor, Phillips' Science of Dental Materials, 11th edition. St. Louis: Saunders/Elsevier Science; 2003

2. Powers JM, Sakaguchi RL (editors).Craig's Restorative Dental Materials, 12th edition. St. Louis: Mosby/Elsevier; 2006

3. Boyer DB, Edie JW.Composition of clinically aged amalgam restorations. Dent Mater 199063146150

4. International Organization for Standardization: ISO 22674. Dentistry- Metallic materials for fixed and removable appliances; 2006.

5. I. Watanabe, M. Atsuta, K. Yasuda, K. Hisatsune, Dimensional changes related to ordering in an AuCu-3wt%Ga alloy at intraoral temperature. Dent Mater 1994106369374

6. American National Standard/American Dental Association Specification No.Metal-ceramic dental restorative systems: 2000 (Reaffirmed 2010This specification is a modified adoption of ISO 9693:1999.

7. American Dental Association, Council on Scientific Affairs.Revised classification system for alloys for fixed prosthodontics. [http://www.ada.org/2190.aspx]

8. S. F. Rosenstiel, M. F. Land, J. Fujimoto, Fixed. Contemporary, 4th. Prosthodontics, Mosby. edition, Mosby/Elsevier; 2006

9. Mackert JR Jr, Ringle RD, Fairhurst CW.High-temperature behavior of a Pd-Ag Alloy for porcelain. J Dent Res 1983621212291235

10. H. Ohno, Y. Kanzawa, Structural changes in the oxidation zones of gold alloys for porcelain bonding containing small amounts of Fe and Sn. J Dent Res 19856416773

11. E. Papazoglou, W. A. Brantley, J. C. Mitchell, Z. Cai, A. B. Carr, high-palladium. New, studies. alloys, the. of, with. interface, porcelain, Int J Prosthodont 199694315322

12. E. Papazoglou, W. A. Brantley, A. B. Carr, W. M. Johnston, Porcelain adherence to high-palladium alloys. J Prosthet Dent 1993705 386394

13. Vermilyea SG, Huget EF, Vilca JM.Observations on gold-palladium-silver and gold-palladium alloys. J Prosthet Dent 1980443294299

14. A. B. Carr, Z. Cai, W. A. Brantley, J. C. Mitchell, New high-palladium casting alloys. Part 2. Effects of heat treatment and burnout temperature. Int J Prosthodont 199363233241

15. S. G. Vermilyea, Z. Cai, W. A. Brantley, J. C. Mitchell, Metallurgical structure and microhardness of four new palladium-based alloys. J Prosthodont 199654288294

16. W. H. Guo, W. A. Brantley, D. Li, W. A. Clark, P. Monaghan, R. H. Heshmati, Annealing study palladium-silver dental alloys: Vickers hardness measurements and SEM microstructural observations. J Mater Sci Mater Med 2007181111118

17. Baran GR.The metallurgy of Ni-Cr alloys for fixed prosthodontics. J Prosthet Dent 1983505639650

18. H. F. Morris, K. Asgar, A. P. Rowe, Nasjleti, The influence of heat treatments on several types of base-metal removable partial denture alloys. J Prosthet Dent 1979414388395

19. Bridgeport DA, Brantley WA, Herman PF.Cobalt-chromium and nickel-chromium alloys for removable prosthodontics, Part 1: Mechanical properties. J Prosthodont 199323144150

20. Grasso. J. E. Waldmeier, G. J. Norburg, Nowak, Bend testing of wrought wire removable partial denture alloys. J Prosthet Dent 1996765559565

21. Titanium. A. Donachie Jr, Guide. Technical, edition. 2nd, Park. O. H. A. S. M. Materials, International; 2000

22. Jeong Y-H, Choe H-C, Brantley WA.Nanostructured thin film formation on femtosecond laser-textured Ti-35Nb-xZr alloy for biomedical applications. Thin Solid Films 20105191546684675

23. H. Lee, S. Dregia, S. Akbar, S. Alhoshan, of. . Growth, D. Ti, O. nanowires, Ti. on, alloys. Ti, oxidation. by, J Nanomaterials 2010DOI:10.1155/2010/503186].

24. W. A. Brantley, T. Eliades, (editors, Materials. Orthodontic, Scientific, Aspects. Clinical, Stuttgart: Thieme; 2001

25. A. Verstrynge, J. Van Humbeeck, G. Willems, In-vitro evaluation of the material characteristics of stainless steel and beta-titanium orthodontic wires. Am J Orthod Dentofacial Orthop 2006130 4460470

26. A. J. Goldberg, R. Vanderby, C. J. Jr Burstone, Reduction in the modulus of elasticity in orthodontic wires. J Dent Res 1977561012271231

27. Asgharnia MK, Brantley WA.Comparison of bending and tension tests for orthodontic wires. Am J Orthod 1986893228236

28. Khier SE, Brantley WA, Fournelle RA.Structure and mechanical properties of as-received and heat-treated stainless steel orthodontic wires. Am J Orthod Dentofacial Orthop 199199310318

29. Backofen WA, Gales GF.Heat treating stainless steel wire for orthodontics. Am J Orthod 1951212117124

30. Funk AC.The heat treatment of stainless steel. Angle Orthod 1951129138

31. Howe GL, Greener EH, Crimmins DS.Mechanical properties and stress relief of stainless steel orthodontic wire. Angle Orthod 1968383244249

32. Fillmore GM, Tomlinson JL.Heat treatment of cobalt-chromium alloys of various tempers. Angle Orthod 1979492126130

33. J. Goldberg, C. J. Burstone, An evaluation of beta titanium alloys for use in orthodontic appliances. J Dent Res 1979582593599

34. C. J. Burstone, A. J. Goldberg, titanium. A. Beta, orthodontic. new, Am. J. alloy, Orthod, 1980772121132

35. Andreasen GF, Hilleman TB. An evaluation of 55 cobalt substituted nitinol wire for use in orthodontics. J Am Dent Assoc 197182613731375

36. G. F. Andreasen, P. R. A. Brady, hypothesis. use, . for, wire. nitinol, orthodontics. for, Angle Orthod 1972422172177

37. Andreasen GF, Morrow RE.Laboratory and clinical analyses of nitinol wire. Am J Orthod 1978732142151

38. C. J. Burstone, B. Qin, J. Y. Morton, Ni. Chinese, wire. Ti-a, orthodontic. new, alloy, Am J Orthod 1985876445452

39. F. Miura, M. Mogi, Y. Ohura, H. Hamanaka, The super-elastic property of the Japanese NiTi alloy wire for use in orthodontics. Am J Orthod Dentofacial Orthop 1986901110

40. Khier SE, Brantley WA Fournelle RA.Bending properties of superelastic and nonsuperelastic nickel-titanium orthodontic wires. Am J Orthod Dentofacial Orthop 1991994310318

41. M. L. Fletcher, S. Miyake, W. A. Brantley, D. S. C. Culbertson, studies. bending, a. of, shape-memory. new, wire. orthodontic, J Dent Res 1992Spec Iss A): 169, Abstract 505

42. T. W. Duerig, K. N. Melton, D. Stöckel, C. M. Wayman, (editors, Engineering Aspects of Shape Memory Alloys. London: Butterworth-Heinemann; 1990

43. R. P. A. Kusy, of. review, archwires. contemporary, properties. Their, characteristics, Angle Orthod 1997633197207

44. F. Miura, M. Mogi, Y. Ohura, Japanese NiTi alloy wire: Use of the direct electric resistance heat treatment method. Eur J Orthod 1988103187191

45. Bradley TG, Brantley WA, Culbertson BM.Differential scanning calorimetry (DSC) analyses of superelastic and nonsuperelastic nickel-titanium orthodontic wires. Am J Orthod Dentofacial Orthop 19961096589597

46. W. A. Brantley, M. Iijima, T. H. Grentzer, D. S. C. Temperature-modulated, new. provides, about. insight, wire. nickel-titanium, transformations, Am J Orthod Dentofacial Orthop 20031244387394

47. International Organization for Standardization: ISO 15841. Dentistry- Wires for use in orthodontics;2006ANSI/ADA Specification 32-OrthodonticWires: 2006 is an identical adoption of this ISO standard.

48. Khier SE. Structural Characterization, Biomechanical Properties, and Potentiodynamic Polarization Behavior of Nickel-Titanium Orthodontic Wire Alloys [Ph.D.Dissertation]. Milwaukee, WI, USA: Marquette University; 1988

49. W. A. Brantley, W. Guo, W. A. Clark, M. Iijima, Microstructural studies of 35°C copper Ni-Ti orthodontic wire and TEM confirmation of low-temperature martensite transformation. Dent Mater 2008242204210

50. C. Canalda-Sahli, E. Brau-Aguadé, J. Sentís-Vilalta, The effect of sterilization on bending and torsional properties of K-files manufactured with different metallic alloys. Int Endod J 19983114852

51. H. Walia, W. A. Brantley, H. Gerstein, An initial investigation of the bending and torsional properties of Nitinol root canal files. J Endod 198814346351

52. Thompson SA.An overview of nickel-titanium alloys used in dentistry. Int Endod J 2000334297310

53. Alapati SB, Brantley WA, Svec TA, PowersJM, Nusstein JM, Daehn GS. SEM observations of nickel-titanium rotary endodontic instruments that fractured during clinical Use. J Endod 20053114043

54. P. Parashos, H. H. Messer, Rotary NiTi instrument fracture and its consequences. J Endod 2006321110311043

55. S. B. Alapati, W. A. Brantley, M. Iijima, S. R. Schricker, J. M. Nusstein, U. Li-M, T. A. Svec, R. D. Micro-X, D. S. C. temperature-modulated, of. investigation, rotary. nickel-titanium, instruments. endodontic, Dent Mater 2009251012211229

56. T. R. Tripi, A. Bonaccorso, E. Rapisarda, V. Tripi, G. G. Condorelli, R. Marino, I. Fragalà, Depositions of nitrogen on NiTi instruments. J Endod 2002287497500

57. S. Zinelis, M. Darabara, T. Takase, K. Ogane, G. D. Papadimitriou, The effect of thermal treatment on the resistance of nickel-titanium rotary files in cyclic fatigue. Oral Surg Oral Med Oral Pathol Oral Radiol Endod 20071036843847

58. S. B. Alapati, W. A. Brantley, M. Iijima, W. A. Clark, L. Kovarik, C. Buie, J. Liu, W. B. Johnson, Metallurgical characterization of a new nickel-titanium wire for rotary endodontic instruments. J Endod 2009351115891593

59. W. A. Brantley, T. A. Svec, M. Iijima, J. M. Powers, T. H. Grentzer, Differential scanning calorimetric studies of nickel titanium rotary endodontic instruments. J Endod 2002288567572

60. J. Liu, Characterization of New Rotary Endodontic Instruments Fabricated from Special Thermomechanically Processed NiTi Wire [Ph.D. Dissertation]. Columbus, OH, USA: Ohio State University; 2009

61. X. Hou, Y. Yahata, Y. Hayashi, A. Ebihara, T. Hanawa, H. Suda, Phase transformation behaviour and bending property of twisted nickel-titanium endodontic instruments. Int Endod J 2011443253258

62. Y. Shen, H. M. Zhou, Y. F. Zheng, L. Campbell, B. Peng, M. Haapasalo, Metallurgical characterization of controlled memory wire nickel-titanium rotary instruments. J Endod 2011371115661571

63. Y. Yahata, T. Yoneyama, Y. Hayashi, A. Ebihara, H. Doi, T. Hanawa, H. Suda, Effect of heat treatment on transformation temperatures and bending properties of nickel-titanium endodontic instruments. Int Endod J 2009427621626

64. K. Otsuka, X. Ren, Physical metallurgy of Ti-Ni-based shape memory alloys. Prog Mater Sci 2005505511678

2

# EFFECT OF PROCESSING PARAMETERS ON THE MECHANICAL PROPERTIES OF HEAVY SECTION DUCTILE IRON

**Mohamed Mahmoud Mourad, Shimaa El-Hadad,**
**Mervat Mohamed Ibrahim, andAdel Abdelmonem Nofal**

Central Metallurgical Research and Development Institute,
P.O. Box 87, Helwan, Cairo, Egypt

## ABSTRACT

The main objective of the current work is to investigate the influence of different inoculation conditions on the microstructure and mechanical properties of heavy section ductile iron (DI) castings. Inoculation treatment was done via one step and double step treatments with different amounts of inoculants. The mechanical properties of the fabricated samples were evaluated. The best inoculation procedure in terms of graphite nodules characteristics and mechanical properties was double inoculation with 0.8% inoculants added at first and 0.2% in the late inoculation step. The presence of Sb in one of the cast alloys controlled the growth of graphite nodules in these heavy section ductile iron castings; however low impact toughness was recorded. The matrix structure of ductile cast iron showed a significant influence not only on the strength and impact properties but also on the fracture mode during testing.

# 1. INTRODUCTION

Wind power is a renewable, predictable, and clean source of energy. Substantial capacity can be built up quickly, offering the energy independence demanded by the world's largest and fastest-growing economies. The majority of wind turbine parts are made out of the challenging ductile iron grade EN-GJS-400-18-LT [1]; examples of DI castings for windmill are shown in Figure 1. Therefore, attention is given for this grade of iron. The production of heavy section ductile iron castings, however, still faces real difficulties, represented in the formation of degenerated graphite morphologies in the ductile iron microstructures, which drastically reduces the impact as well as fatigue properties [2–4]. The requirements for high strength and high impact properties in DI at low temperatures for some applications as windmill are very strict. The European Standard EN-GJS-400-18U-LT commonly referred to as GGG 40.3 is an example. This standard not only has the normal mechanical requirements for ferritic iron but also specifies V-notched Charpy impact requirements of 12 J at −20°C and minimum elongation of 18% [5]. In order to achieve these mechanical properties, the microstructure should be well controlled in terms of nodularity, nodule count, and matrix phases. This is because high nodule count impairs impact strength and low nodule count results in intercellular brittle phases which are detrimental to strength and ductility; see Figure 2. A minimum nodularity of 95% is critical; poorly shaped nodules act as stress riser and as initiation site for fracture under impact. Also, a matrix which is fully ferritic and free of cell boundary phases is important to satisfy the mechanical properties of windmill castings [4, 6].

(a) (b)

**Figure 1.** Ductile iron wind mill components, (a) hubs, (b) shaft.

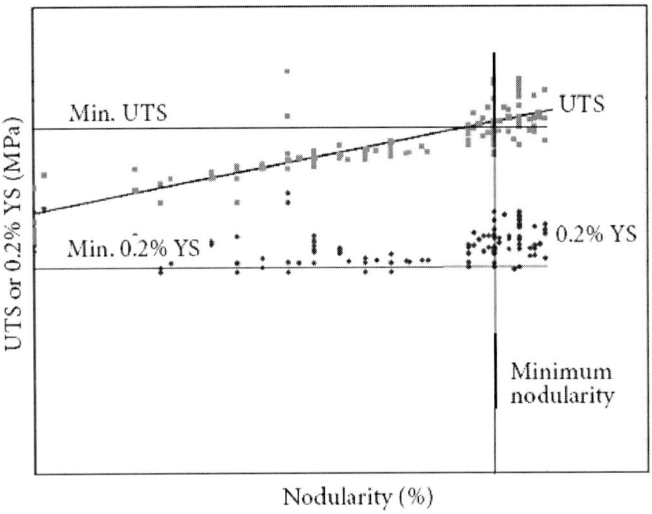


**Figure 2.** A plot representing UTS and YS, showing a drop in the property as the nodularity is reduced [9].

Due to the increasing interest on sound heavy section DI castings and the controversial arguments about the foundry practice to control their microstructure (with particular reference to inoculation), this work is aimed at supporting the research about the influence of inoculation on microstructure and mechanical properties of heavy section DI castings. Though the section thickness in the current work is not comparable to that of the windmill, the experiments and the results of the present work give attention to some important points to be taken into consideration for casting of heavier sections of DI.

# 2. EXPERIMENTAL WORK

## 2.1. Casting

A sand mould of cavity shaped as block of 400 mm length and (100 mm × 100 mm) cross section was prepared. The moulds were made completely of sodium silicate bonded sand. A 100 kg charge, which consists of 80% sorel metal, 20% steel scrap, and the required amounts of carburizer and ferrosilicon alloy, was melted in a medium frequency induction furnace of 100 KW, 100 Kg capacity. The molten metal was poured at 1520°C. The Mg treatment to obtain nodular graphite was performed using Vortex method with Fe-Si-Mg master alloy [7]. The molten metal was inoculated by foundry grade Fe-Si. Three different alloys were obtained depending on the inoculation method: alloy A was inoculated via one step with all the amount of inoculants added in the vortex, 0.5%. Alloy B was inoculated in two steps; in the first step, half of Fe-Si quantity (0.25%) was added in the vortex and the rest was added in the ladle which is known as (in-stream/in-mold) inoculation as the second step. In case of alloy C, 0.8% of the inoculants was added in the first step while 0.2% was added in the late inoculation step.

Though Sb is not a common additive in windmill DI castings, it is recommended in heavy section ductile iron castings because of its positive effect on graphite degeneration problem [8]. Therefore, another alloy D (double step, 0.5% inoculants) was cast where 0.02% of Sb was added intentionally to know the influence of Sb as a graphite antidegenerating element on the applicability of a DI alloy for windmill. The chemical composition of the four alloys expressed in mass content of the alloying elements is shown in Table 1.

**Table 1.** Chemical composition of DCI alloys.

| Alloy number | Chemical composition, Wt % | | | | | | |
|---|---|---|---|---|---|---|---|
| | C | Si | Mn | P | S | Mg | Sb |
| A | 3.52 | 2.69 | 0.193 | 0.032 | 0.009 | 0.060 | — |
| B | 3.60 | 2.69 | 0.22 | 0.033 | 0.012 | 0.062 | — |
| C | 3.51 | 2.16 | 0.19 | 0.0232 | 0.0148 | 0.0421 | — |
| D | 3.69 | 2.58 | 0.309 | 0.029 | 0.006 | 0.044 | 0.02 |

In order to control the molten metal processing, thermal analysis was performed using the standard equipment necessary that consists of sand cup instrumented with a thermocouple connected to a microcomputer. The thermocouples were placed in the middle part of the cup. The obtained cooling curves were used as a guide during the casting process.

## 2.2. Microstructure Investigation

The samples were cut of the produced blocks A, B, C, and D for metallographic examination. The quantitative analysis was performed for each sample and the average values of graphite nodularity and nodule count were considered. This was done using image analysis software.

## 2.3. Mechanical Testing

The tensile test was performed using a tensile test machine at a loading rate of 10 mm/min. Three tensile test samples shaped according to (ASTM E8) were used for each alloy of A–D. The average values of the yield strengths, tensile strengths, and elongations were calculated. After the test, the topography of the fracture surface was investigated using scanning electron microscopy. An instrumented impact test machine using notched samples was used. The impact strength was evaluated according to ASTM-E23. An average of five readings was considered.

# 3. RESULTS AND DISCUSSION

## 3.1. Microstructure of As-Cast Samples

The microstructures of the as-cast samples A, B, C, and D before and after etching are shown, respectively, in Figures 3 and 4. In all the micrographs shown in Figure 3, homogeneity of graphite in the matrices is observed. After etching with nital, they present different microstructures depending on the pearlite content in the matrix, Figures 4(a)–4(d). The corresponding quantitative analysis of graphite nodules count and nodularity is summarized in Table 2.

The one step inoculation (A), Figure 3(a), resulted in a considerably low nodularity of about 70% and nodule count of 40%. The amount of pearlite in the microstructure was around 2%, Figure 4(a). Using the double step inoculation (0.5% inoculant), alloy B did not show notable change in nodule count or in the nodularity. This is a disadvantage considering the high nodule count required for some applications like wind mill (not more than 100–200/mm$^2$). Following the recommended double step inoculation from the industrial expertise [7], 0.8% inoculants were added in the first step and 0.2% at the mold. In this case, a significant improvement in the graphite nodularity of about 95% was achieved and the nodule count was increased from 40 to 80/mm$^2$. On the other hand, alloy D containing 0.02% Sb could achieve the best nodularity of 98% and nodule count was almost the same as alloy C.

**Table 2.** Microstructure description of the DCI alloys.

| Alloy | Microstructure description |
|---|---|
| A | Matrix 98% ferrite, 2% pearlite<br>Nodularity 70%, nodule count 40/mm$^2$ |
| B | Matrix 95% ferrite, 5% pearlite<br>Nodularity around 71%, nodule count 40/mm$^2$ |
| C | Almost fully ferritic matrix<br>Nodularity around 95%, nodule count 80/mm$^2$ |
| D | Matrix 2% ferrite, 98% pearlite<br>Nodularity 98%, nodule count 80/mm$^2$ |

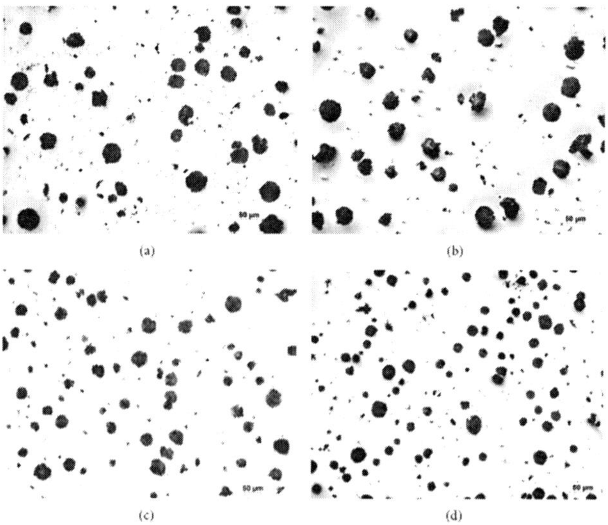


**Figure 3.** Optical micrographs (before etching) of the four DI alloys (a) A, (b) B, (c) C, and (d) D.

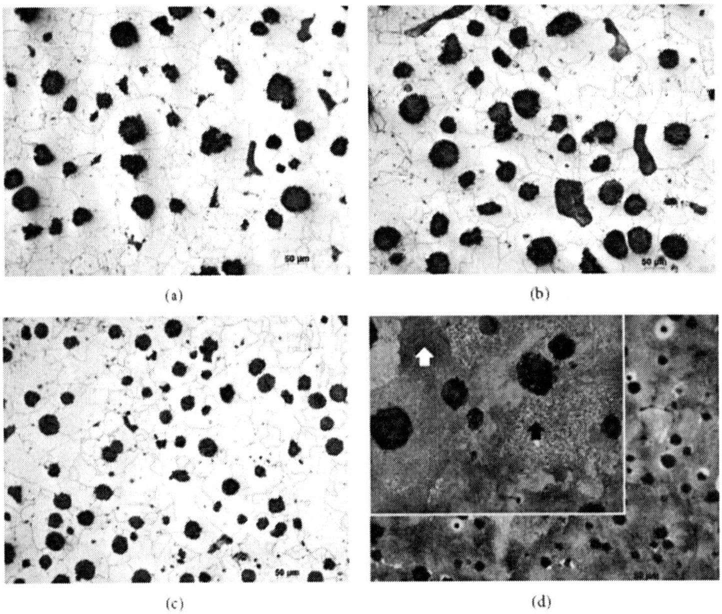

**Figure 4.** Optical micrographs (after etching) of the four DI alloys (a) A, (b) B, (c) C, and (d) D.

Referring to Zhe et al. [8, 11], appropriate content of antimony addition plays a significant role in increasing the nucleation rate and inhibiting the deterioration of graphite nodules in heavy section ductile iron castings. This is due to the fact that Sb surrounds the graphite nodule and prevents the carbon diffusion from the matrix to the nodules as shown in the line scan and elemental map of Figure 5. However, in the presence of Sb, the matrix is almost fully pearlitic; containing fine and coarse pearlite lamellas. This percentage of pearlite is detrimental in terms of the subzero impact toughness required for this application, especially in the cold regions. Comparing Figures 3(a) and 3(b) to Figures 3(c) and 3(d), it is clear that the size of graphite nodules was increased in alloys A and B which resulted in relatively lower nodule counts. This can be further confirmed by the nodules size distribution carried out by image analyzer and represented in Figure 6. According to the nodules size classification [9] for ductile iron casting, shown in Figure 7, the largest nodule takes grade 3 (25–50 mm) while the smallest one (less than 1.5 mm) is graded as 8. In case of alloys A and B, the particles of grade 5 (6–12 mm) represented the majority among all sizes. On the other hand, most of the nodules in alloys C and D were of grade 6 (3–6 mm). When the nodules combine together, their size increased and their count accordingly decreased.

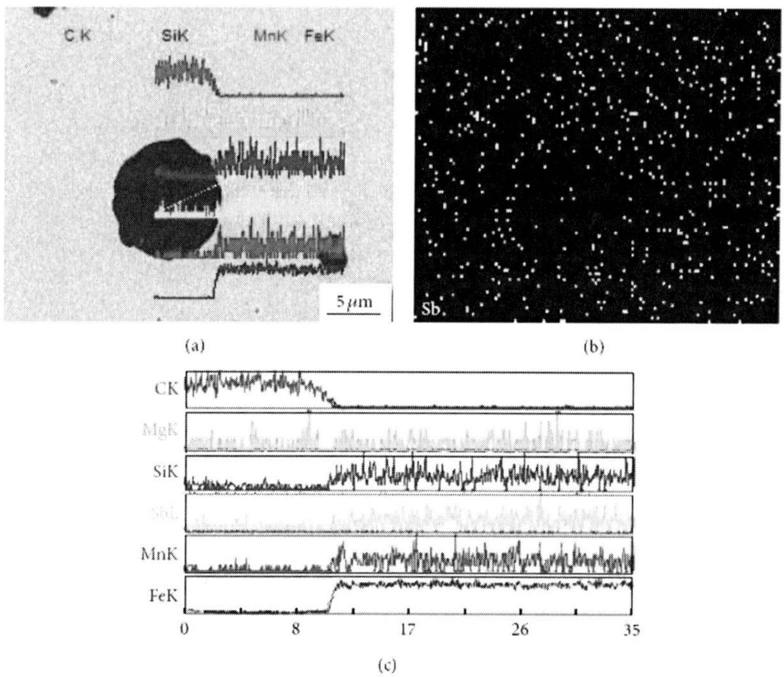


**Figure 5.** Line scan and elemental map of alloy D sample showing the distribution of Sb around a graphite nodule.

## 3.2. Thermal Analysis

Thermal analysis was performed in order to compare the solidification behavior of the melt after the different inoculation conditions. Figures 8 and 9 are magnified parts of the cooling curves showing the eutectic and eutectoid reactions, respectively, in the different alloys. Some characteristic temperatures calculated from the cooling curves by means of the first time-derivative are shown in Table 3. The characteristic temperatures are denoted by $T_{eu}$ for the eutectic temperature (undercooling) and $T_{eu}$ for the maximum bulk eutectic temperature (recalescence). Recalescence rate was also calculated as $T_{eu}$ divided by $T_{eu}$. Comparing $T_{eu}$ and $T_{eu}$ values of Figures 8(a)–8(d), it is clear that the different inoculation procedures influenced both of the eutectic undercooling and the recalescence temperatures. The one step inoculation, alloy (A), showed the lowest $T_{eu}$ and $T_{eu}$ as presented in Table 3. On the other hand, the double step inoculation of alloys B and D using 0.5% inoculant increased $T_{eu}$ and $T_{eu}$ by 7 degrees in case of alloy B. The presence of antimony in alloy D caused further increase in both of the temperatures up to 13 degrees as shown in Table 3. Double inoculation with larger amount of inoculant (1%), alloy C, showed lower $T_{eu}$ and $T_{eu}$ in comparison to alloy B. According to Ferroa et al. [12], the second solidification step (starting from $T_{eu}$ ) refers to the nucleation of secondary nodules and growth of the corresponding eutectic spheroidal graphite cells. Based on it, the double inoculation process can increase $T_{eu}$ and $T_{eu}$ and this effect is decreased when larger amount of inoculant is added. The highest $T_{eu}$ and $T_{eu}$ recorded for Sb containing alloy are owed to the ability of Sb to support formation of typical nodules in heavy section ductile iron [8].

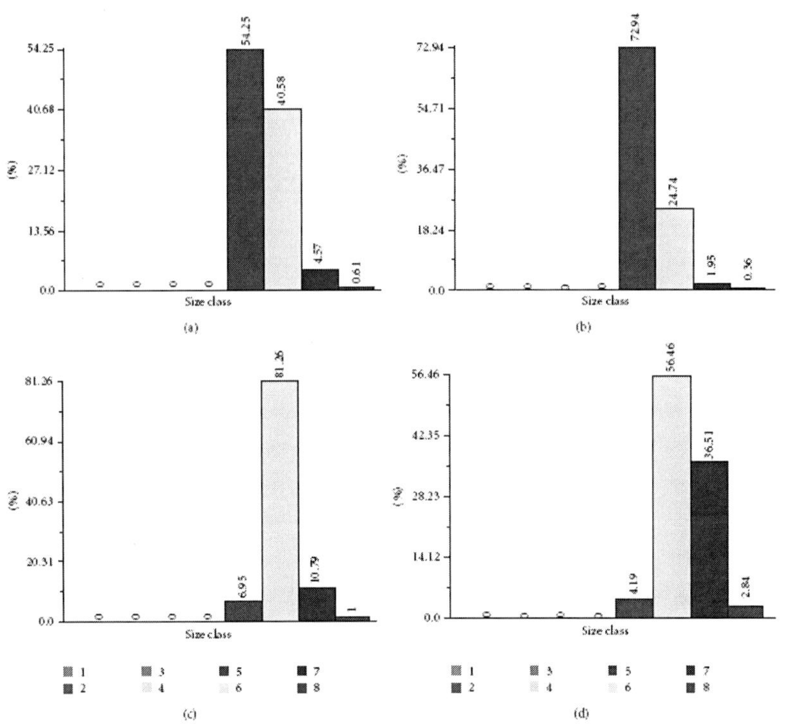


**Figure 6.** Distribution of particles size in alloys A, B, C, and D, respectively.

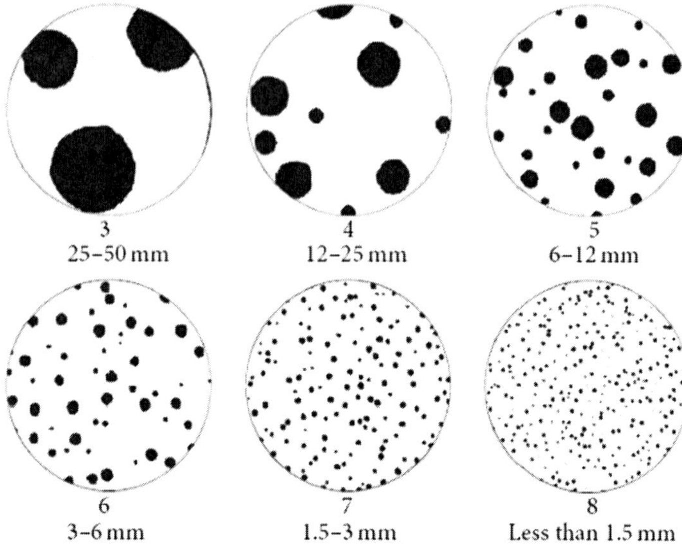


**Figure 7.** Graphite size classes in ductile iron [10]. Diagrammatic representations of the standard graphite nodule sizes in ductile iron at ×100 magnification (from VI).

**Table 3.** Cooling curve characteristics temperatures.

| Alloy | $T_{eu}$ | $T_{Re}$ |
|---|---|---|
| A | 1140 | 1143 |
| B | 1147 | 1150 |
| C | 1142.5 | 1152 |
| D | 1153 | 1157.5 |

Figure 9 represents the eutectoid part of the cooling curves of alloys A, B, C, and D. It is clear that the eutectoid reaction in alloy D is different than that of the other alloys. The increased eutectoid undercooling denoted as dTin Figure 9 is due to the formation of pearlite phase. It is known that Sb is a pearlite stabilizer [4] since it limits the diffusion of carbon from pearlite to graphite nodules as represented in the line scan of Figure 5.

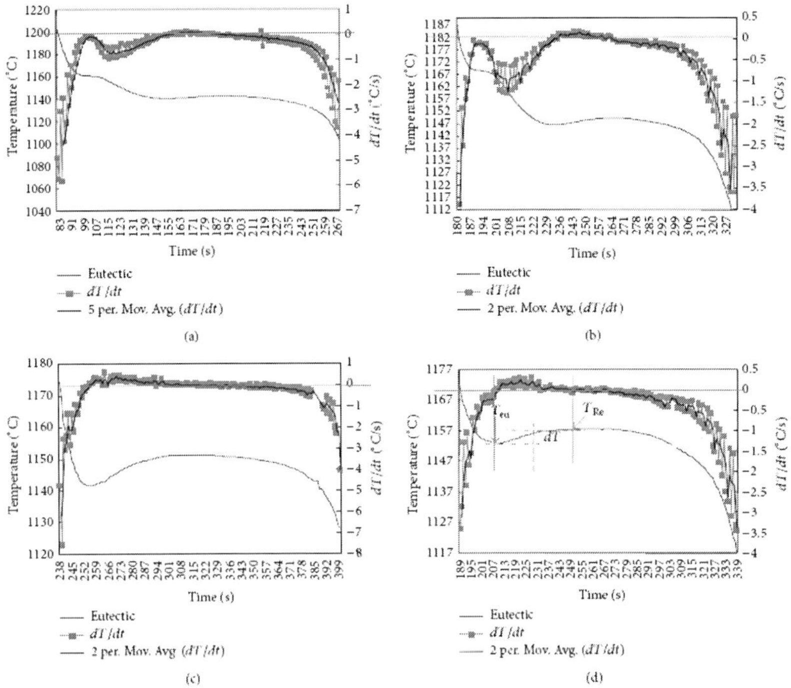


**Figure 8.** Magnified part of the cooling curves showing the change in the amount of eutectic undercooling of the DI alloys (a) A, (b) B, (c) C, and (d) D.

## 3.3. Mechanical Properties Evaluation

### 3.3.1. Tensile Test

A considerably high yield strength and tensile strength with a minimum of 240 MPa and 400 MPa, respectively, are required along with 18% elongation according to European Standard EN-GJS-400-18U-LT for windmill DCI. The results of tensile test are shown in Figure 10 for A–D samples. It is clear that there is no evident difference in the strength values of alloy A (one step inoculation, 0.5% inoculant) and alloy B (double step, 0.5% inoculant). Both of these alloys could meet the standard values of yield and tensile strength with an elongation exceeding 20%. Alloy C (double step, 1% inoculants) with almost ferritic matrix showed relatively lower ultimate and yield strength values, while higher elongation of 24% was recorded. This increased elongation is due to the relative absence of pearlite. Though this alloy presented better graphite nodules characteristics compared to alloys A and B, Table 2, the strength decreased relatively in (double step, 1% inoculant) alloy C. A good yield, tensile strength, and fracture toughness are expected in the microstructure presenting 100% of ferrite due to the presence of silicon and manganese. Both elements harden and strengthen the ferrite and accordingly increase the yield and tensile strength [10]. However, in the current experiments, the unintended decrease in the Si content to 2.16, Table 1, occurred. This can explain the unexpected reduction in the strength of alloy C.

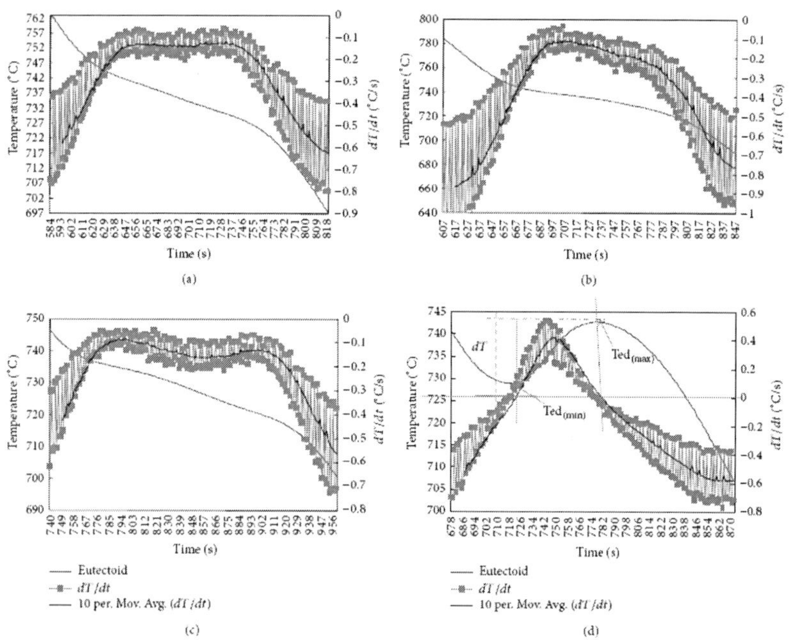


**Figure 9.** Magnified part of the cooling curves showing the change in the amount of eutectoid undercooling of the DI alloys (a) A, (b) B, (c) C, and (d) D.

In case of alloy D, a significant increase in both of the yield and tensile strength was observed, Figure 10. The matrix in this microstructure, Figure 4(d), is almost pearlitic. It is well known that pearlite hardens the matrix and, at the same time, increased pearlite content enhances the resistance of the matrix [13]. In addition, presence of antimony (around 0.02%) improves the graphite morphology and plays a significant role in increasing the graphite nucleation rate and inhibiting the deterioration of graphite nodules in heavy section ductile iron castings [14]. Therefore, the best nodularity (98%) could be obtained in alloy D containing Sb, Table 2.

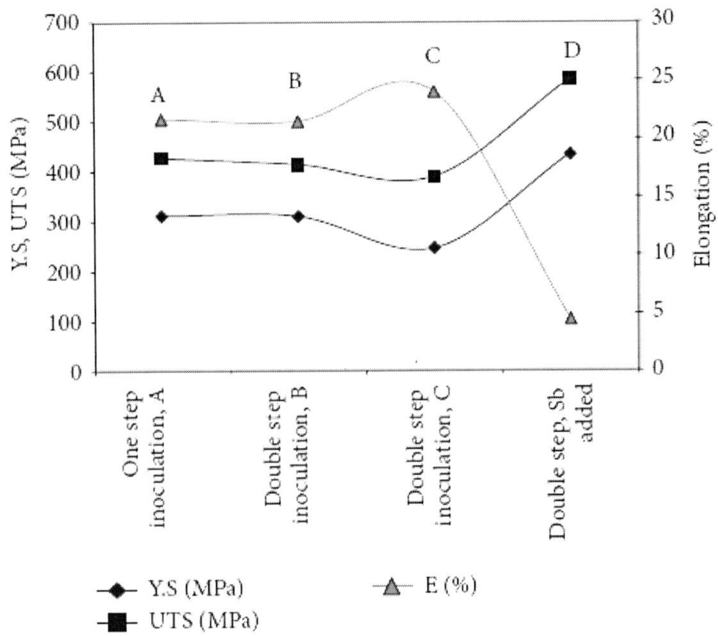


**Figure 10.** Tensile test results of the DI alloys (a) A, (b) B, (c) C, and D.

## 3.3.2. Failure Surface Analysis

The fracture surfaces of the tensile test samples were firstly analyzed with the naked eye. We observed that the fracture surface of alloy D is almost bright while that of alloys A, B, and C is relatively dark as shown in Figure11. Both of these zones were then observed with a field emission microscopy and represented in Figure 12. A deep analysis of Figures 12(a)–12(h) indicates that different mechanisms of fracture have occurred. This difference in behavior of fracture was expected due to the difference in the matrix structure. Alloys A, B, and C have ferritic matrix, while matrix of alloy D is pearlitic, Figures 4(a)–4(d). It was reported that the fracture mechanism is much related to the ferrite and pearlite content in as-cast ductile iron [15]. In the samples with ferrite matrix, the surface is composed of many fracture dimples. These small dimples observed in the fracture surface represent voids coalescence as shown in the magnified micrographs of Figures 12(b), 12(d), and 12(f). According to Figures 12(a)–12(f), there are different failure surfaces, indicating that failure mode seems to be ductile-brittle intergranular.

**Figure 11.** Macrostructure of the fracture surface of tensile samples of alloys (a) A, (b) B, (c) C, and (d) D.

The topography of the sample, Figures 12(g) and 12(h), shows that small plastic deformation is observed. This deformation has to do with the brittle fracture that can be identified when observing the surface that fails. They revealed that cleavage has occurred in brittle fracture mode by cleavage (intragranular rupture). Concluding, the matrix structure of ductile cast iron always affects not only the strength and impact properties but also the fracture mode during testing.

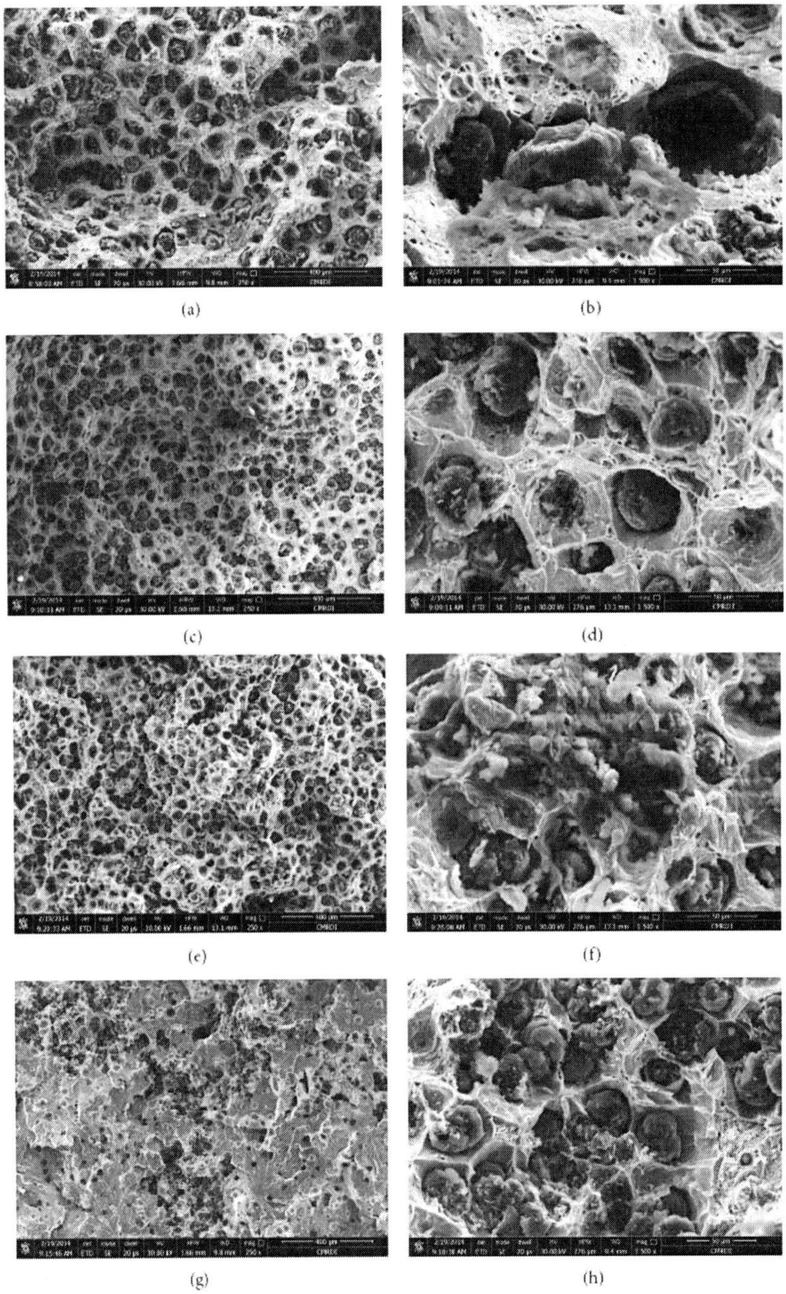

**Figure 12.** Topography of the fractured tensile test samples at low and high magnification of alloy A ((a), (b)), alloy B ((c), (d)), alloy C ((e), (f)), and alloy D ((g), (h)).

### 3.3.3. Impact Toughness

In this paper, the study is focused on a heavy section ferritic ductile iron. The ferrite (alpha iron) has a good ductility (plasticity) and a very good ability to resist the impact energy at low temperatures. The presence of a small quantity of residual pearlite in the ferrite matrix may change the tensile properties and resilience of ductile iron. It increases material hardness and reduces the impact energy necessary to failure [15]. Also the impact properties of ferritic cast iron are affected by the quantity of graphite nodules and the nodularity; as the nodules become larger in size or its count increases, the mechanical properties decrease accordingly [16]. Based on it, we have two factors that influence the impact properties in the current study: the ferrite content and the graphite nodule features.

The results of impact toughness are shown in Figure 13 for the four tested samples. The impact toughness recorded for alloy A (one step inoculation, 0.5%) was around 16 J and increased to 19 J for alloy B (double step inoculation, 0.5%). Here it is worth noting that the effect of increased graphite nodularity in alloy B compared to A (70–80%) was translated to higher impact toughness. Though the pearlite content increased from 2 to 5% from alloy A to B, respectively, the improvement in graphite nodularity was more effective. Chaengkham and Srichandr [17] reported that the more the form of the graphite deviates from the ideal spherical shape, the more the ductility and strength decrease. This is due to stress concentration caused by the nonspherical nodules. In alloy C (fully ferritic matrix, 95% nodularity), the impact value increased to around 25 J, which was expected due to the ferritic matrix along with well rounded nodules. Considering the standard impact properties for windmill (12 J at −20°C), the three samples with ferrite matrix could achieve the standard. In contrast, sample D with almost pearlitic matrix showed very low impact energy of about 6 J. Though this alloy has the highest nodularity (~98%), the presence of pearlitic matrix which is known to reduce the fracture toughness [10] was more effective. Therefore, the difficult challenge in the production of DI alloys for windmill is to obtain fully ferritic matrix while achieving the standard strength conditions [6].

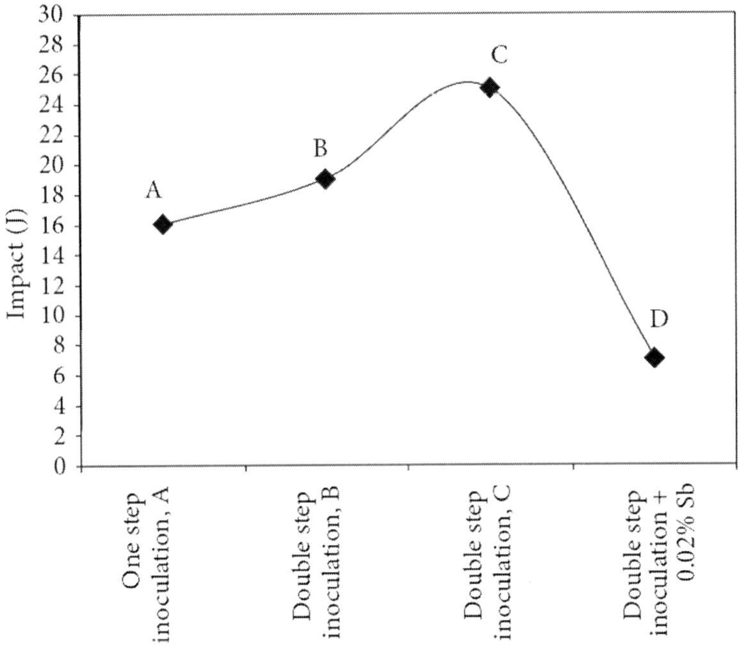


**Figure 13.** Impact toughness of alloys A, B, C, and D.

## 4. CONCLUSIONS

In this study, some heavy section ductile cast iron castings were prepared in order to study the influence of different inoculation methods on microstructure and mechanical properties of these alloys. The following conclusions were derived.(i)The best inoculation procedures in the current experiments in terms of graphite nodules characteristics and mechanical properties are double inoculation with 0.8% inoculants added at first and 0.2% in the late inoculation step.(ii)In antimony containing alloy, Sb showed a positive influence on controlling the growth of graphite nodules in the heavy section DCI castings; however low impact toughness was recorded.(iii)The matrix structure of ductile cast iron showed a significant influence not only on the strength and impact properties but also on the fracture mode during testing.

# REFERENCES

1. M. J. Fallon, "Experiences in the manufacture of ductile irons," The Foundryman, vol. 88, no. 9, pp. 308–318, 1995.

2. G. K. Bouse, J. Parolini, N. Rojek, and C. Zhou, "Quality requirements for general electric ductile iron castings exceeding 10T used for wind and gas turbine components," in Proceedings of the 69th World Foundry Congress (WFC '10), vol. 3, Hangzhou, China, 2010.

3. C. M. Ecob, A Review of Common Metallurgical Defects in Ductile Cast Iron, Elkem Foundry Products Division, Oslo, Norway, 2004.

4. I. Riposan, M. Chisamera, S. Stan, and T. Skaland, "Surface Graphite Degeneration in Ductile Iron Castings for Resin Molds," Tsinghua Science and Technology, vol. 13, no. 2, pp. 157–163, 2008.

5. Rio Tinto Iron & Titanium, Ductile Iron Data for Design Engineers, Rio Tinto Iron & Titanium, Montreal, Canada, 1990.

6. H. Roedter and M. Gagne, "Ductile iron for heavy section windmill castings," in Proceedings of the Keith Millis Symposium on Ductile Cast Iron, pp. 176–182, 2003.

7. H. Bakkerus and B. J. van der Holst, "Dutch develop vortex process for ductile iron treatment," Modern Casting, vol. 71, no. 3, pp. 41–44, 1981.

8. L. Zhe, C. Weiping, and D. Yu, "Influence of cooling rate and antimony addition content on graphite morphology and mechanical properties of a ductile iron," China Foundry, vol. 9, no. 2, pp. 114–118, 2012.

9. http://www.foundry.elkem.com/.

10. A. Alhussein, M. Risbet, A. Bastien, J. P. Chobaut, D. Balloy, and J. Favergeon, "Influence of silicon and addition elements on the mechanical behavior of ferritic ductile cast iron," Materials Science and Engineering A, vol. 605, pp. 222–228, 2014.

11. L. Zhenhua and L. Yanxiang, "Effect of Sb on the graphite morphology and mechanical properties in heavy section ductile iron," Materials Science Forum, vol. 475–479, pp. 2769–2772, 2005.

12. P. Ferroa, A. Fabrizi, R. Cervob, and C. Carollo, "Effect of inoculant containing rare earth metals and bismuth on microstructure and mechanical properties of heavy-section near-eutectic ductile iron castings," Journal of Materials Processing Technology, vol. 213, no. 9, pp. 1601–1608, 2013.

13. R. A. Gonzaga, P. Martínez Landa, A. Perez, and P. Villanueva, "Mechanical properties dependency of the pearlite content of ductile irons," Journal of Achievements in Materials and Manufacturing Engineering, vol. 33, no. 2, pp. 150–158, 2009.

14. R. A. Gonzaga, "Influence of ferrite and pearlite content on mechanical properties of ductile cast irons,"Materials Science and Engineering: A, vol. 567, pp. 1–8, 2013.

15. W. L. Bradley and M. N. Srinivasan, "Fracture and fracture toughness of cast irons," International Materials Reviews, vol. 35, no. 1, pp. 129–161, 1990.

16. M. Shirani and G. Härkegård, "Large scale axial fatigue testing of ductile cast iron for heavy section wind turbine components," Engineering Failure Analysis, vol. 18, no. 6, pp. 1496–1510, 2011.

17. P. Chaengkham and P. Srichandr, "Continuously cast ductile iron: processing, structures, and properties,"Journal of Materials Processing Technology, vol. 211, no. 8, pp. 1372–1378, 2011.

3

# MICROSTRUCTURE-PROPERTY RELATIONSHIP IN ADVANCED NI-BASED SUPERALLOYS

**Hiroto Kitaguchi**[1]

[1] Department of Materials, University of Oxford, OX1 3PH, Oxford, UK

## 1. INTRODUCTION

Ni based superalloys have been developed more or less empirically over the past 60 years from a simple Ni-Cr matrix to the present multi element and phase systems[1], having a fully austenitic face centred cubic (fcc) structure which maintains a superior tensile, fatigue and creep properties at high temperature to a body centred cubic (bcc) alloy[1]. One of the major applications of Ni superalloys is gas turbine engines. They comprise over 50% of the weight of advanced aircraft engines and include wrought and cast turbine blades and powder metallurgy (P/M) route turbine discs [1].

One of the most important goals of engine design is increasing turbine entry temperature (TET): the temperature of the hot gases entering the turbine arrangement [2]. This implies that the resistance against the environmental attack, i.e. high temperature, under a severe mechanical force is the priority challenge and indeed Ni based superalloys are used in the hottest as well as the highest tensile pressure of the gas turbine engine component as shown in the schematic diagram in Fig. 1. Nowadays, for the advanced cast single crystal superalloys in the turbine blades, the alloy capability exceeds 1,000ºC [2]. In this chapter, the polycrystalline Ni superalloys, which have slightly less temperature capability up to 800°C, applied in the turbine discs and the adjoined shafts, will be introduced focusing on their microstructures correlating with the mechanical properties.

# 2. MICROSTRUCTURE (SECOND PHASES)

## 2.1. Hardening Precipitates

From the point of view of microstructure, Ni superalloys are complex [4]. The fcc matrix, known as γ, mainly consists of nickel, cobalt, iron, chromium and molybdenum. The strength of superalloys are conferred by the hardening precipitates known as γ′ ($Ni_3Al$ based $L1_2$ structure) (Fig. 2). In some nickel – iron superalloys such as IN718 and IN706, which contain niobium, they are hardened by γ′′ ($Ni_3Nb$ based $D0_{22}$ structure) (Fig. 3) [2]. Homogeneously distributed coherent hardening precipitates confer excellent tensile and fatigue life properties at high temperatures. Their volume fraction is controlled by the nominal chemical composition. The size and the morphology are controlled by the process and their crystallographic relations with γ matrix. The precipitates arise close to the solvus temperature grow larger which subsequently restrict the grain growth pinning grain boundaries (Fig. 4). On the other hand, the precipitates arise at lower temperature such as during cooling after heat treatment stay small (Fig. 4 (left hand side of the image)). γ′ has the perfect coherency with the γ matrix, hence their morphologies are mostly sphere, whereas γ′′ has a tall crystal unit tetragonal structure where *a* axis has the identical lattice parameter with the γ matrix but *c* axis has nearly double the length of the γ, hence γ′′ always precipitate with the perfect coherency on the basal plane with γ and grow along the longitudinal direction (Fig. 5).

## 2.2. Carbides and Borides

Carbon and boron are added as a grain boundary strengthener by segregating in the grain boundaries and forming carbides and borides. They are believed to be formed during solidification, aging treatment which strengthen grain boundaries at elevated temperatures but the ones arising during service must be controlled carefully since they can impair properties [4].

Carbides are traditionally classified by their chemical composition, mainly MC, $M_6C$ and $M_{23}C_6$, where M stands for metal elements such as Ti, Cr, Nb, Mo, Hf and Ta [4].

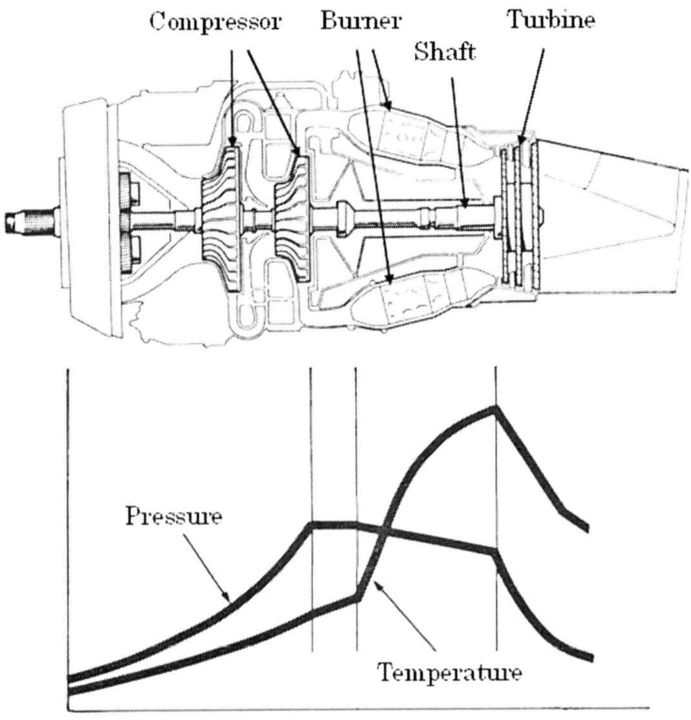


**Figure 1.** a) Schematic diagram of a turbine engine Ref. [**3**]. (b) Schematic diagram of the temperature and pressure gradients throughout the engine component correlating with the diagram (a) Ref. [**4**]

MC carbides are usually coarse (Fig. 6), having a fcc densely packed structure [4]. Ti, Nb, Hf and Ta are the main metal elements. They are very strong and are normally considered to be some of the most stable compounds in nature, justified by their high precipitation and melting temperature: they are believed to precipitate during processing shortly after solidification of the superalloy [4]. They usually have little or no orientation relationship with the alloy matrix [4].

$M_6C$ carbides have a complex cubic structure and they precipitate when the alloy contains highly refractory elements, for example Mo and W. These carbides are believed to be the product of MC carbide decomposition during service or relatively high heat treatment between

815 and 980ºC [4]. The examples of the micrographs of $M_6C$ can be found in Ref. [5, 6].

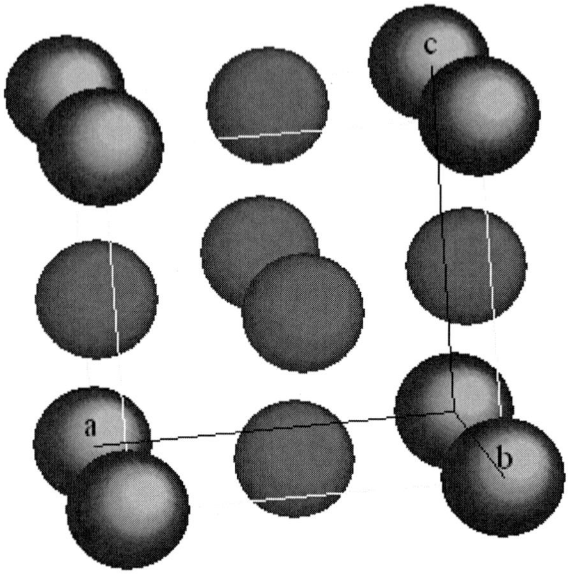


**Figure 2.** $\gamma'$ $L1_2$ structure. Ni atoms are blue and Al purple

$M_{23}C_6$ carbides (Fig. 7) form mainly along grain boundaries at a relatively low temperature for carbides: between 760 and 980ºC. The crystal structure is complex cubic structure. The lattice parameter is exactly three times larger than $\gamma$ matrix, hence they precipitate with cube-cube orientation with the matrix (Fig. 8). They are believed to form either by the decomposition of MC or $M_6C$ or they nucleate directly on the grain boundaries. They are known as having a high content of Cr. $M_{23}C_6$carbides have a significant effect on Ni based superalloy properties [4] since they are profuse in alloys with moderate to high Cr content [4] and are controversial carbides. Firstly, this is because their different morphologies: the blocky shaped ones at grain boundaries have a beneficial effect on rupture strength; on the contrary the film ones are regarded as promoting early rupture failure [4]. Secondly, this is because that they make a Cr depleted zone (Fig. 9) around the precipitate. In this area, it is difficult to form a protective oxide, namely $Cr_2O_3$, due to lack of Cr.

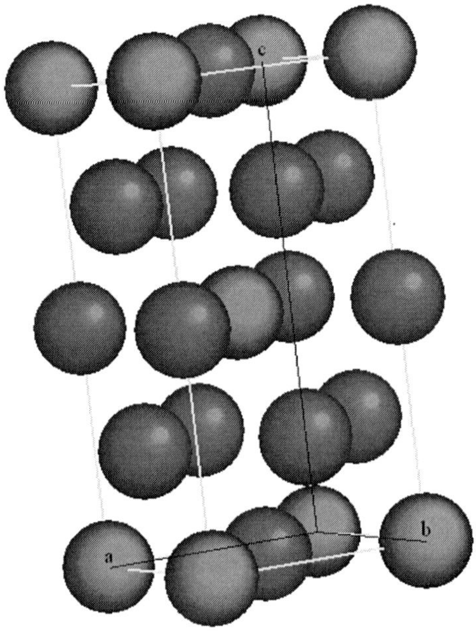


**Figure 3.** $\gamma''$ $D0_{22}$ structure. Ni atoms are blue and Nb, Al and Ti purple

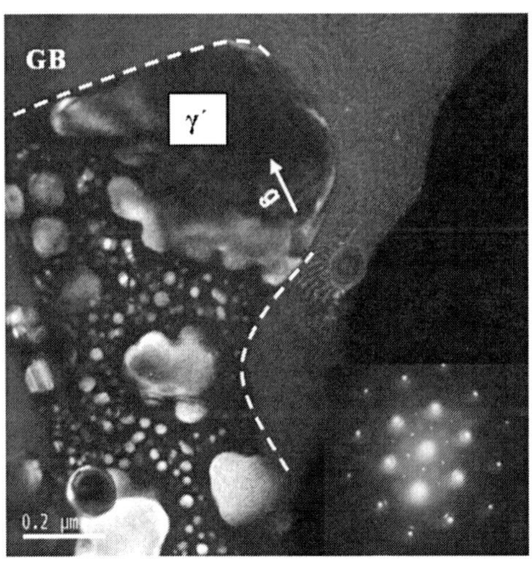


**Figure 4.** TEM dark field (DF) image. $\gamma'$ pinning grain boundary, shown by the white dashed line. The small spherically shaped precipitates inside the grain are also $\gamma'$. (g = 01 $1^{-}1^{-}$ B = [111])

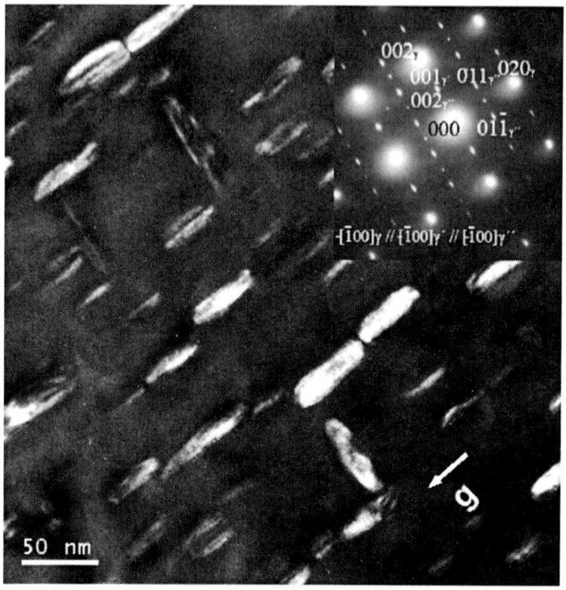


**Figure 5.** TEM DF image of the γ″ in IN718. The growth direction is c axis parallel to the a axis of γ ( ($g_{\gamma''}$ = 0 0 2) B = [**100**]).

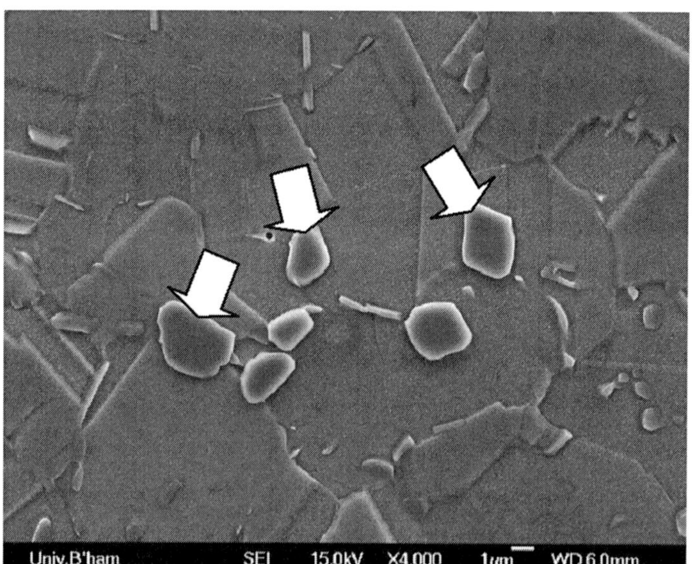


**Figure 6.** Coarse Nb and Ti based carbide in IN718

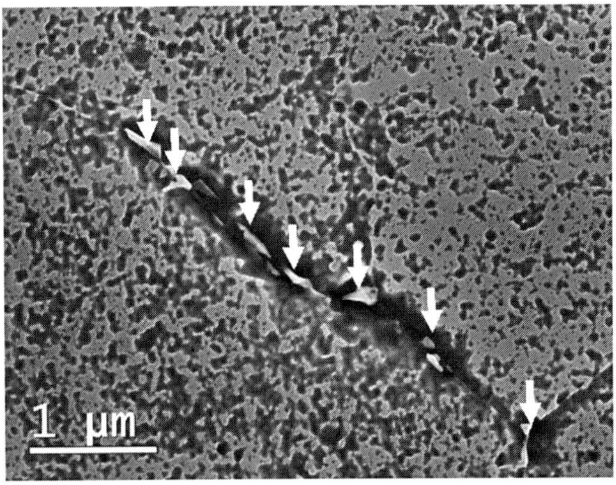


**Figure 7.** Fine $M_{23}C_6$ type carbides precipitate along the grain boundary running diagonally.

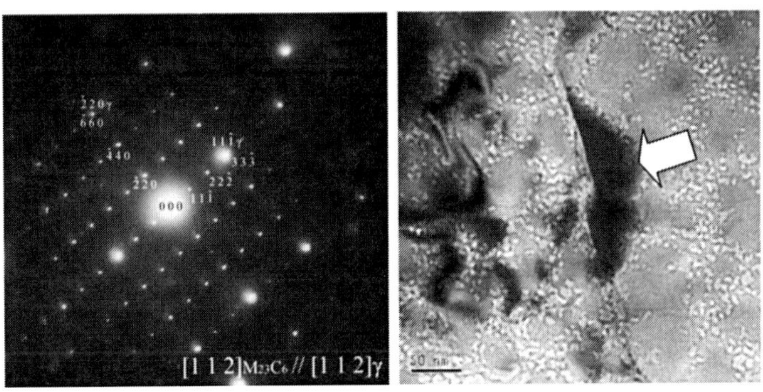


**Figure 8.** $M_{23}C_6$ and $\gamma$ matrix perfect coherent diffraction pattern (left) and the bright field image from another beam direction to make $M_{23}C_6$ outstanding (right)

It is broadly acknowledged that boron segregation along the grain boundary increases the cohesive strength of the grain boundaries. The role of borides is, however, still under open discussion. Those so far identified have a base centred tetragonal (BCT), $M_3B_2$ [4] or $M_5B_3$ [7] formula, where M is typically a refractory element, namely Mo or Cr. They appear as various shapes such as blocky to half-moon [4]. The examples shown in Fig. 10 were found in an advanced polycrystalline Ni superalloy after a thermal exposure at 980°C for 1 hour.

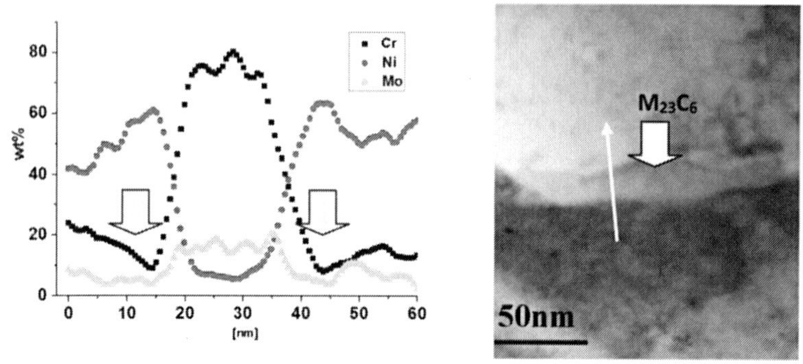


**Figure 9.** Left: STEM EDX line scan results across $M_{23}C_6$ revealed the Cr depletion as indicated by the arrows (Cr nominal composition is 15 wt%). Right: STEM bright field image of the $M_{23}C_6$ (the thin arrow shows the length and the direction of the EDX line scan)

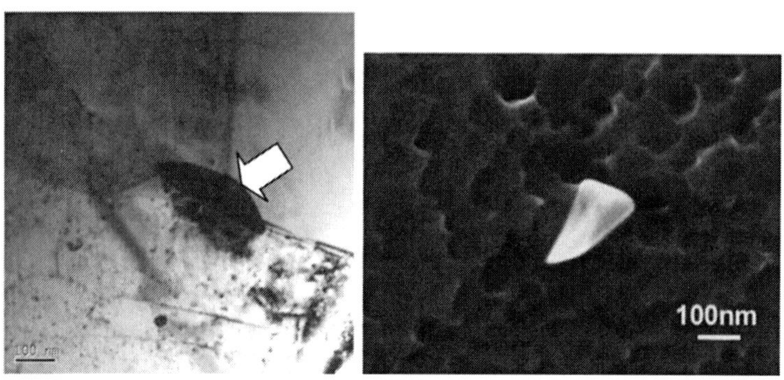


**Figure 10.** Some examples of $M_5B_3$ type boride appeared in TEM bright field (left) and in SEM (right).

## 2.3. Other Phases

Adding excess quantity of refractory elements, such as Mo, W and Re, promotes the precipitation of hard intermetallic phases [2], so called

TCP phase, which are believed to deteriorate the alloy ductility [4] and the creep life [8]. In the ternary phase diagrams for superalloy elements, such as Ni-Cr-Mo, there are two phase spaces: one is austenite (γ) fcc and the other is bcc [4]. Between these two major fields, a band of numerous small phase volumes can be identified such as σ, μ, R and so on [4], which are characterized firstly as having a high and uniform packing density of atoms[2] and secondly as having complex crystal structures [2], either hcp, body centred tetragonal or rhombohedral. With the careful control of these refractory elements, TCP phases occur after a long time service or a prolonged heat treatment [9]. Some are believed to be the products of transformation from another beneficial phase: for example η($Ni_3X$) results from γ′ [4] and σ has the same crystal structure as that of $M_{23}C_6$, but without the carbon atoms. The example of σ phase shown in Fig. 11 was found to be Cr, Mo and Co based chemistry after a thermal exposure at 720°C for 1,100 hours in a newly developed advanced Ni superalloy. The second phases introduced above and some other important second phases for the Ni superalloy microstructure are summarized in Table 1.

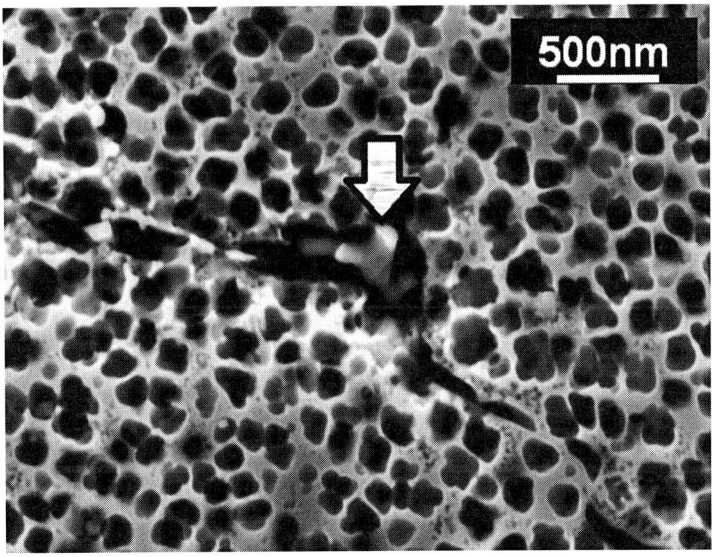


**Figure 11.** Sigma (σ) phase precipitates on the grain boundary running diagonally from top left to bottom right

**Table 1.** Summary of second phases in the polycrystalline Ni based superalloys [**10**] The lattice parameter may vary (less than 5%) by changing chemical composition.

| Phase | Prototype | Pearson symbol | Strukturbericht symbol | Lattice [nm] | Chemical Composition (Appx) |
|---|---|---|---|---|---|
| γ' | $Cu_3Au$ | cP4 | $L1_2$ | a 0.36 | $(Ni\ Co)_3(Al\ Ti)$ |
| γ'' | $Al_3Ti$ | tI8 | $D0_{22}$ | a 0.36 c 0.74 | $(Ni\ Fe)_3(Nb\ Al\ Ti)$ |
| MC | NaCl | cF8 | B1 | a 0.44 | (Ti Ta)C or TiC, TaC, NbC, WC |
| $M_6C$ | $Fe_3W_3C$ | cF112 | $E9_3$ | a 1.11 | $(Mo\ Cr\ W)_6C$ |
| $M_7C_3$ | $Cr_7C_3$ | oP40 | $D10_1$ | a 0.45 b 0.70 c 1.21 | $Cr_7C_3$ |
| $M_{23}C_6$ | $Cr_{23}C_6$ | cF116 | $D8_4$ | a 1.07 | $Cr_{21}Mo_2C_6$ |
| $M_5B_3$ | $Cr_5B_3$ | tI32 | $D8_1$ | a 0.55 c 1.06 | $(Cr\ Mo)_5B_3$ |
| $M_3B_2$ | $Si_2U_3$ | tP10 | $D5_a$ | a 0.60 c 0.32 | $(Mo\ Cr)_3B_2$ |
| σ | CrNi | tP30 | $D8_b$ | a 0.88 c 0.46 | Cr Mo Co based |
| δ | $Cu_3Ti$ (β) | oP8 | $D0_a$ | a 0.51 b 0.43 c 0.46 | $Ni_3Nb$ |
| η | $Ni_3Ti$ | hP16 | $D0_{24}$ | a 0.51 c 0.83 | $Ni_3(Ti\ Ta)$ |
| μ | $Fe_7W_6$ | hR13 | $D8_5$ | a 0.48 c 2.5 | Mo Co based |

# 3. MICROSTRUCTURES AND MECHANICAL PROPERTIES

It is worth noting the microstructure related mechanical properties in detail. We will discuss briefly how microstructure affects various mechanical properties in polycrystalline Ni superalloys.

Altering grain sizes results in various effects with regard to the different mechanical properties. Tensile and fatigue life properties are optimized by a fine grain microstructure, on the other hand, good creep and fatigue crack growth properties at elevated temperature are favoured by a coarse grain microstructure [2]. The former is a result of grain orientation and stress concentration by dislocation movement along the slip plane [2]. The latter is about intergranular crack propagation susceptibility. For example, Bain *et al* [11] showed the significance of the grain size for the crack growth rate using UDIMET720. Testing at 650°C, the crack growth rate reduced by more than two orders of magnitude by changing the size from 20 to 350 µm in diameter. (Fig. 12).

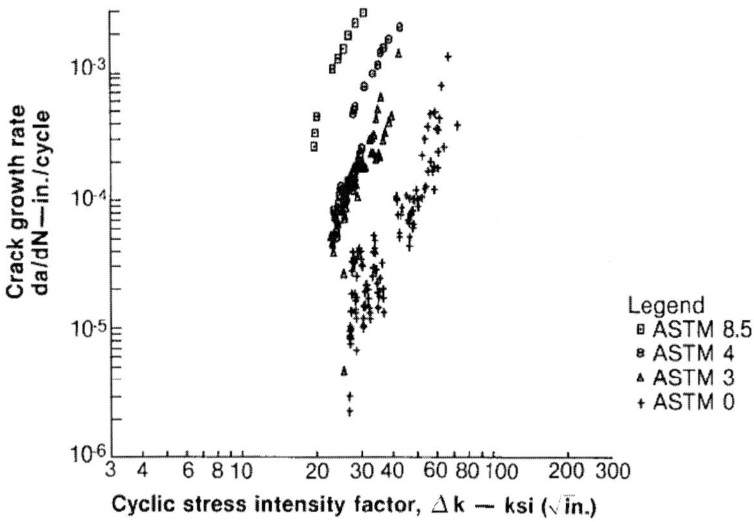


**Figure 12.** UDIMET 720 fatigue crack growth rate for different grain sizes (ASTM grain size between 0 and 8.5: 360µm and 19µm in diameter) tested at 650ºC [**11**]

The size of the hardening precipitates significantly affects the yield strength of the material via the interaction between the precipitate and the dislocation. If the precipitates are large, dislocation bowing around the precipitates becomes dominant; for small sized precipitates, dislocation cutting becomes dominant.

For bowing

$$\tau = \frac{G * b}{L - 2r} \tag{1}$$

and for cutting

$$\tau = \frac{r * \gamma * \pi}{b * L} \tag{2}$$

$\tau$ is the strength of the material, $G$ is the shear modulus, $b$ is the magnitude of the Burgers vector, $L$ is the distance between the hardening precipitates, $r$ is the radius of the precipitates and $\gamma$ is the surface energy. In general in Ni-Al binary system, the optimum size to maximize the strength is found to be around 5 - 30 nm in diameter (Fig. 13). The size of the precipitates also affects the creep strain as shown in Fig. 14. In their study [12], the size of the precipitate was changed by changing the heat treatment temperature and time and found that the smaller the precipitate the slower the creep strain rate is, which is achieved via the smaller $\gamma'$ - $\gamma'$ channel width [12, 13].

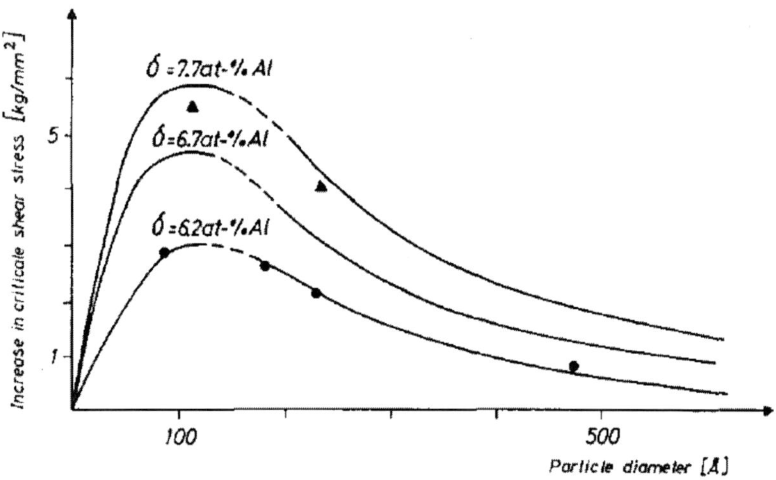


**Figure 13.** $\gamma'$ particle diameter against the critical shear stress in Ni-Al system [**28**]

It is well known fact that in general both the yield strength and the creep rupture strength increases by increasing the hardening precipitate volume fraction [2]. Historically, low cycle fatigue life was the main concern for turbine disc alloys, but fatigue crack growth rate and damage tolerant design have attracted more attention over the last two decades [11, 14]. They can be strongly influenced not only by the size of the grains as introduced above, but also by the size of the precipitates; the striking results were shown in Ref. [15, 16]. The results show that the larger the hardening precipitates the better the crack growth property. However, this conflicts with the creep life property as mentioned above. Research on damage tolerant design originally started to investigate the grain boundary chemistry since fast crack growth (FCG) is always observed with intergranular cracks and tends to disappear at low temperature. Additionally, transgranular ductile cracking replaces intergranular crack when the tests carried out in the reduced oxygen partial pressure [17, 18] (Fig. 15) Thus, FCG embrittlement has been attributed to oxidation [11, 19]. Grain boundary engineering has been explored by changing the morphology of the grain boundary. For example, Ref. [15, 20] reported a complex grain boundary geometry, so called 'serrated' (Fig. 16), by slow cooling after solution treatment. The result showed slower intergranular crack growth rate than with a normal grain boundary [15]. However, the improvement above did not account for the property change by the different size of the hardening precipitate mentioned above. The fast intergranular crack growth at high temperature in superalloys added a new dimension after intensive studies with regard to the correlation between the hardening precipitate distribution and the crack growth rate. Ref. [15, 16, 21] claimed that the prevention of stress relaxation of the crack tip by the hardening precipitates can increase the crack growth rate. Some experimental work support the idea, for example Andieu *et al* [22] carried out a unique dwell fatigue crack propagation test where oxygen was introduced in different phases of the low cycle fatigue crack growth test and found that it is potent for the fast crack growth when oxygen is introduced at the beginning of the loading rather than introducing in the later part of the loading. This may imply that the oxidation at the crack tip happens during the stress concentrated at the crack tip. Molins et al [23, 24] concluded that the local microstructure at the crack tip, which can be controlled by an appropriate heat treatment against the stress accumulation, can significantly affect the crack propagation behaviour in Ni superalloys. This conclusion recalls an arguable grain boundary microstructure feature, namely the precipitate free zone (PFZ). One suggested that the PFZ would promote plastic deformation and fracture [25, 26]. Another suggested that the PFZ in some nickel alloys is beneficial for crack tip stress relaxation [27].

These findings above suggest that not only the macroscopic structure such as the grain size and the distribution of the hardening precipitates, but also the microscopic structure, such as the grain boundary shape and the relationship with the hardening precipitates, can significantly affect the mechanical properties.

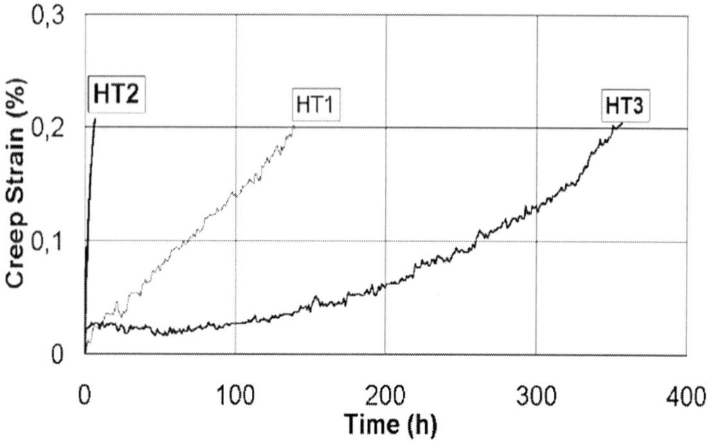


**Figure 14.** Creep strain tested at 700ºC for different heat treatments (HT1, HT2 and HT3) The size of γ′: HT2>HT1>HT3 [**12**].

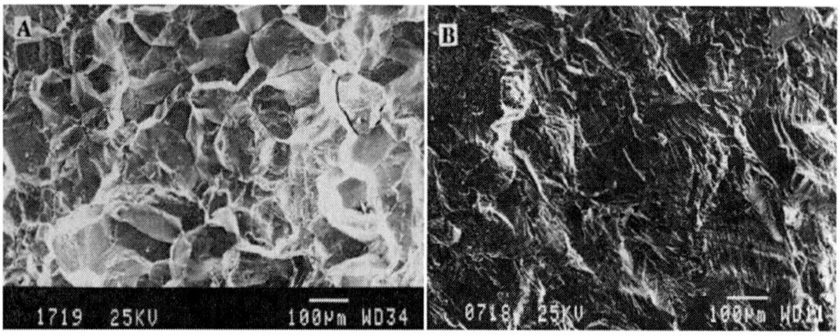


**Figure 15.** Typical intergranular (left) and transgranular (right) fracture surfaces. Alloy 718 tested at 650ºC in air (left) and vacuum (right) [**18**]

# 4. POLYCRYSTALLINE SUPERALLOY GRAIN BOUNDARY STRUCTURE

The details of the Ni superalloy grain boundary microstructure will be demonstrated in this section. Particular attention will be paid to the relationship between the hardening precipitates and the high grain boundaries. Fig. 17 shows the STEM bright field image of the grain boundary and the hardening precipitate morphology in an advanced polycrystalline superalloy. The grain boundary running top left to bottom right cuts through $\gamma'$. This was confirmed by the conventional TEM image analysis combining with the crystallographic analysis that the either side of the $\gamma'$ keeps the coherency with the matrix (Fig. 18). With respect to the morphologies of $\gamma'$ on the grain boundaries, it is the same as those inside the grains. It has, however, two different crystallographic orientations keeping the coherency with the either side of the matrix. This morphology is believed to form during the process with the high boundary mobility [29]. There are at least four different possibilities of interactions between the migrating grain boundaries and the precipitates, which are illustrated in Fig. 19. Following Fig. 19,

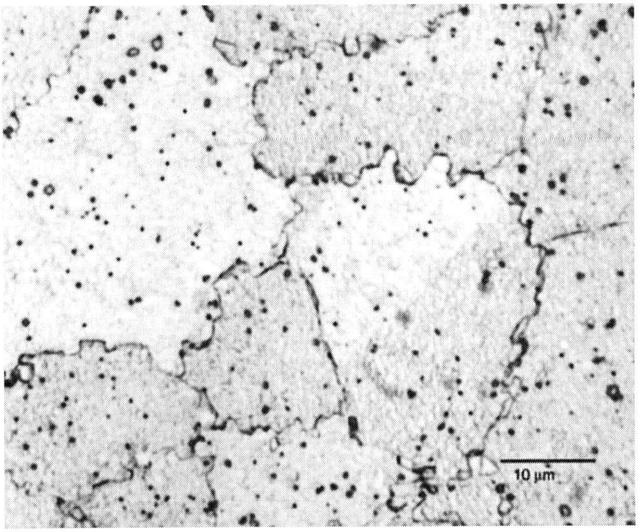


**Figure 16.** Optical microscopy image of serrated grain boundaries. The arrows indicate the serrated boundaries [**15**]

a) the boundary migrates with no effect on the precipitates; the precipitates thus become incoherent after the migrating grain boundary passes through them.

b) the precipitates dissolve in contact with migrating boundary and reprecipitate coherently within the new grain.

c) the grain boundary is held by the coherent precipitates which then coarsen, leading to complete halting of the boundary movement.

d) the grain boundary can pass through the coherent precipitate which undergoes the same orientation change as the grain surrounding it and thereby retains the coherent low-energy interface between the precipitate and the matrix.

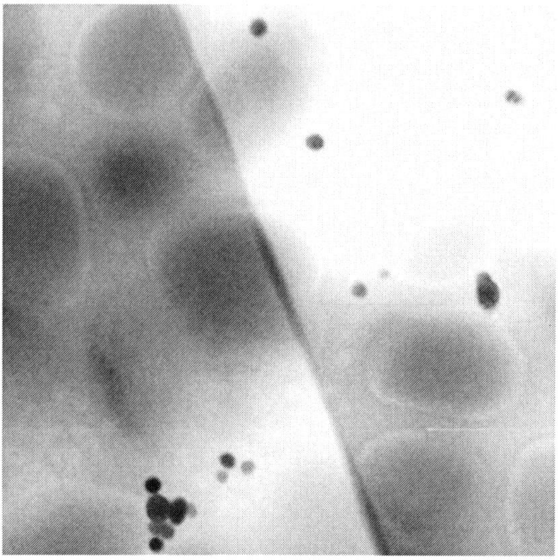

**Figure 17.** General aspect of the high angle grain boundary and γ′ (dark spheres). The grain boundary is running diagonally from top left to bottom right

With regard to theory a), this is often observed with high solvus temperature precipitates, such as carbides and oxides. b) is not applicable in this study. This can be, however, the case for less γ′ volume fraction superalloys with small amount of nominal Al content such as Nimonic PE16 or the case in high temperature very close to the γ′ solvus. With regard to c), it can be applicable in the case of the larger γ′ such as the one in Fig. 4. d) is relevant to explain the results of Fig. 17 and Fig. 18. As the grain boundary impinges on the γ′, the grain boundary apparently cuts off γ′. Firstly, the interface free energy between γ′ and the grain boundary increases. This results in dissolving the γ′ at the interface but due to the supersaturation of γ′ formers, such as Al and Ti, γ′ immediately re - nucleates in the next grain coherent with the next grain discontinuously [30, 31]. Thus, this phenomenon can be concluded the result of γ′ dissolution and subsequent (discontinuous) precipitation.

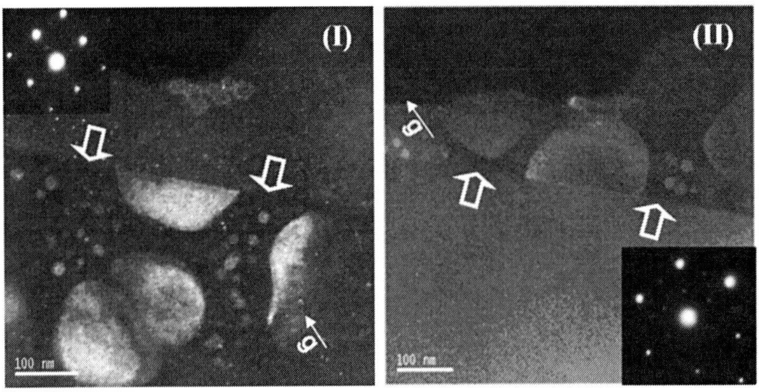


**Figure 18.** A crystallographic analysis of the cutting γ′. The dark field images of the lower and the upper grain are in (I) and (II), respectively. The white arrows indicate the grain boundary

Another example is forming a precipitate free zone (PFZ) as shown in Fig. 20 in IN718. γ′ and γ′′ coexist in IN718, however, γ′′ denude along the grain boundary and form a γ′′ PFZ. On the contrary, the minor hardening precipitate in IN718: γ′ exist along the grain boundary. Vacancy deficiency is one of the causes of the PFZ along grain boundaries due to lack of the nucleation sites as grain boundary acts as a good vacancy sink [32]. One of the important factors to create the PFZ in the superalloys can be the difference of the interfacial free energy, i.e. the free energy between γ′-matrix and γ′′-matrix. The γ′′-γ′′ nucleus channel distance along the grain boundary can be larger than the critical distance to aggregate two γ′′ nuclei. On the other hand, the critical distance for the γ′ is smaller than that of γ′′ or γ′ can nucleate their precipitate independently as γ′ has smaller interfacial energy. Thus, the γ′ nuclei can grow and form precipitates along the grain boundary but not for γ′′ and the γ′′ PFZ arises.

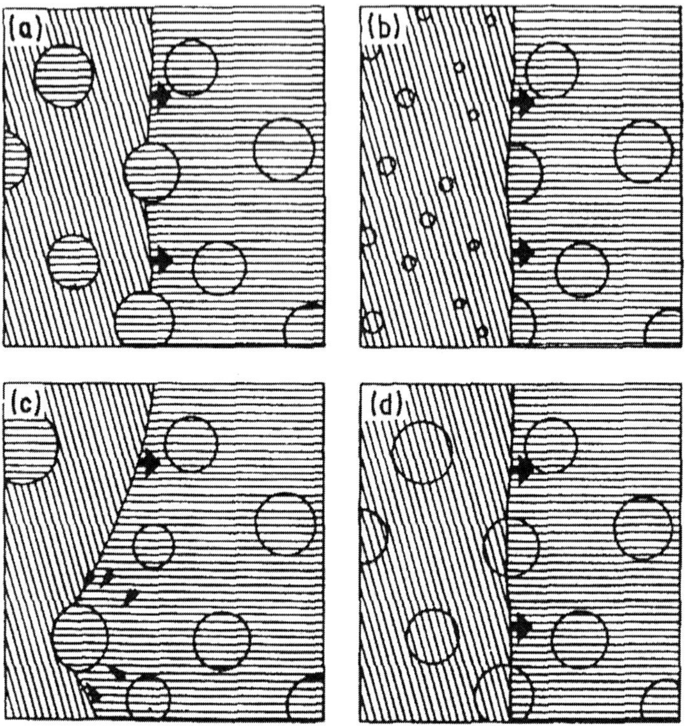


**Figure 19.** The possible interactions between the migrating grain boundary and the precipitate [**29**]

# 5. HIGH TEMPERATURE OXIDATION ALONG GRAIN BOUNDARIES

The context of 'High temperature' in this section is the temperature range of 600-700°C which is the high temperature regime of the disc in the turbine engine application.

It is more than a half century ago, the investigation of the oxidation assisted fast crack growth started. Cr is believed to be an important element for the oxidation assisted crack growth. For example, as shown in Fig. 21, crack growth tests were conducted under various oxygen partial pressure on Ni-Cr binary alloys with 5, 20 and 30 wt% of Cr. It was found that the higher the Cr content the higher the transition oxygen partial pressure from transgranular to intergranular cracking is.

The highest Cr content alloy: Ni-30wt%Cr did not show a transition pressure. Oxidation process on a freshly exposed alloy surface had been characterised intensively and well understood. As illustrated in Fig. 22, both Ni and Cr oxide formation takes place at the beginning of the oxidation [33]. This is particularly important for alloys on the borderline between protective and non-protective behaviour [34]. But in the early stage, the fast kinetics Ni oxide grows quickly and dominates the oxide. In general, there are two types of oxidation: the cation diffusion type and the anion diffusion type [34]. The difference between the two is the movement of the ions; the former involves cation (metal ion) transport, the latter anion (oxygen ion) transport. For the cation diffusion type the oxides form between the oxide and the free surface, but, for the anion diffusion, the oxides form between the oxide and the metal interface. During the transient stage which corresponds to the middle of the illustrations in Fig. 22, $Cr_2O_3$ particles are embedded inside the NiO layer. As NiO grows and the oxidation rate becomes slower, $Cr_2O_3$ soon establishes its own layer underneath the NiO layer, which implies that $Cr_2O_3$ can be the anion diffusion. Eventually the $Cr_2O_3$ layer is completed, where the layer prevents further diffusion of oxygen into the alloy, called passivation [34].

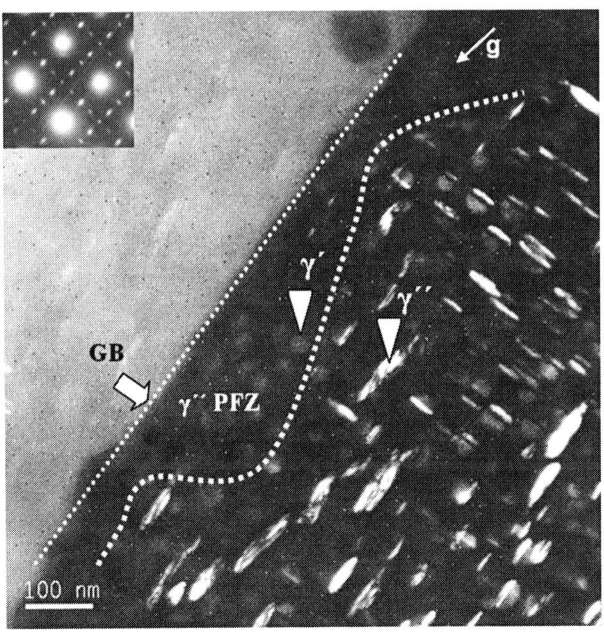


**Figure 20.** γ'' PFZ appeared in IN718, although γ' still exist in the γ''PFZ adjacent to the grain boundary

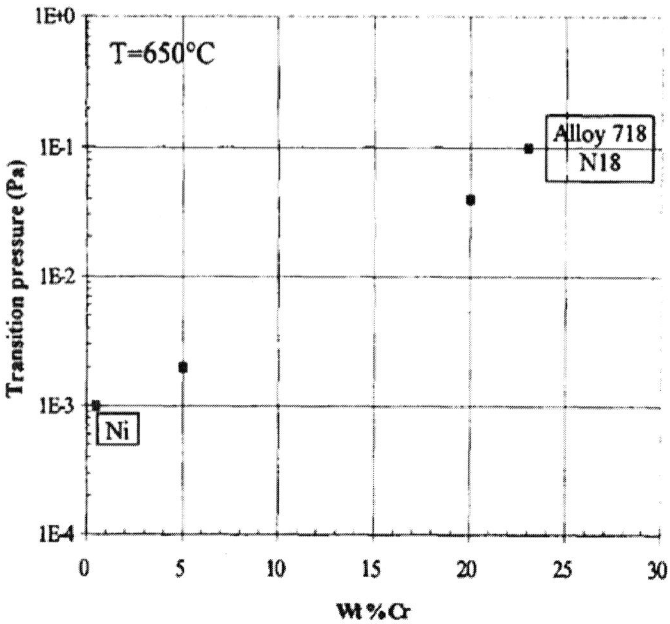


**Figure 21.** Transition oxygen partial pressure from the transgranular to intergranular cracking against the Cr concentration in Ni alloys after crack propagation tests at 650°C [**24**]

Nevertheless, the oxidation process at (ahead) of the crack tip has remained under debate. This is due to the technical difficulties of studying the microstructure of such small regions, which is predictable taking consideration of the size of the crack tip: less than a micron and even smaller for the oxides adjacent. To overcome the problem, so called 'site specific specimen preparation' has been developed since the late 90's using focused ion beam (FIB) technique [35, 36]. The use of gallium ions in a focused ion beam accelerated in a FIB apparatus to energies up to 30 keV enables us to mill specimens selectively to reveal structural features and to deposit films at selected locations. An example carried out in the University of Birmingham UK (2009) is shown in Fig. 23. The specimen is an advanced Ni based superalloy after an interrupted crack propagation test at 650°C in air. A plan-view crack tip TEM specimen was prepared [37]. Using the two different modes: the tungsten deposition to protect the region of interest (Fig. 23 a)) and the milling (trenching) (Fig. 23 b)), the crack tip was transported to a TEM copper grid (c) and d)). The size of the focused Ga ion beam can be achieved as small as a few nanometres in radius. It is possible to prepare the site specific TEM specimen foil as thin as 50 nm in thickness.

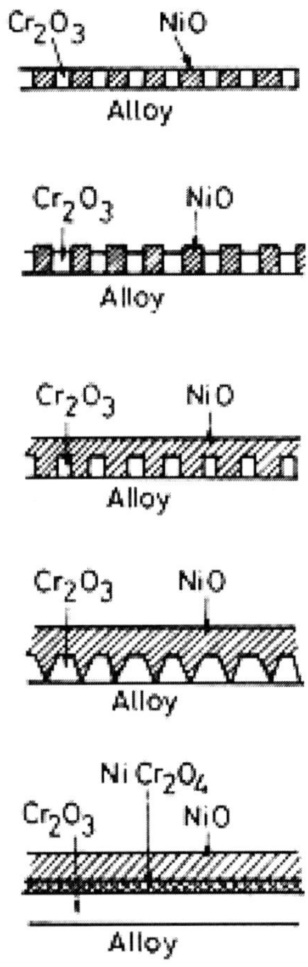


**Figure 22.** Schematic diagram of the oxidation process (from top to bottom) of the freshly exposed Ni alloy [**33**]

Fig. 24 shows TEM bright field images; they are from the same material: polycrystalline advanced superalloy, but they are after different testing conditions. Fig. 24 (a) is from a specimen after interrupting a crack propagation test at 650°C; the intergranular crack propagation was identified. Fig. 24 (b) is from a specimen after interrupting the same crack propagation test mentioned above except for the crack growth rate: 0 μm/s (~0.9 $K_{th}$: just below the crack growth threshold) held for 5 hours. The oxides ahead of the crack tip are also along grain boundaries. The dashed line boxes indicate the area analysed by EDX shown later in this section. Comparing the two bright field images in Fig. 24, it is apparent that the oxides penetration in the metal ahead of the crack tip is approximately 5 times longer for the static crack specimen (b).

The EDX mapping and the EDX line scan of the oxides close to the crack tip for the moving crack specimen from the region I in Fig. 24 are shown in Fig. 25. It is apparent that the grain boundary is completely filled with oxides (oxygen map). There is a Co and Ni rich oxide in the middle. There are Cr rich oxide areas on both sides of the oxide. Cr thus forms a thin layer between the Ni (Co) oxides and the alloy. EDX line scans across the oxide revealed that Ti, Al are also segregating in this region. Crystallographic analysis using selected aperture diffraction confirmed that the middle oxide is cubic $(Ni_x\,Co_{1-x})O$ and the rim oxide is hcp $(Cr\,Al\,Ti)_2O_3$. The higher oxygen partial pressure region in the middle of the oxide is consisted by the Ni and Co oxide. The rim of the oxide between the Ni (Co) oxide and the matrix are consisted by the passive Cr, Al and Ti layer. This is correlating with the freshly exposed Ni alloy oxidation process described above. Fig. 26 shows the oxide chemistry at the tip of the oxide corresponding to the region II in Fig. 24. According to the EDX mapping, the oxides formation manner looks similar to that of the region I; Ni(Co) oxide in the middle and the Cr, Al and Ti oxide in the rim. It is questionable to argue the stoichiometry of the oxide from the results of the EDX cross section line scan due to the x-ray emission from the matrix, however it revealed from the line scan in Fig. 26 that the oxide is Cr based; Ni, Co, Al and Ti deplete in the very tip of the oxide. Fig. 27 shows the chemistry of the oxides in the middle of the oxide ahead of the crack tip for the static crack corresponding to the region x in Fig. 24. EDX mapping revealed that the Ni and Co are depleted even in the middle of the oxide and the line scan revealed that the oxide is consisted mainly by Cr, Al.Fig. 28 shows the tip of the oxide for the static crack corresponding to the region z in Fig. 24. Cr still exists in the oxide in the middle, but particularly at the very tip of the oxide approximately 100 nm or so, Cr is depleted and only Al and Ti enriched at the tip of the oxide (see also the cross section line scan). It is also difficult in this case due to the thickness effect to discuss the stoichiometry of the oxide, however, in this region, Al and Ti based oxide formation takes place.

Table 2 shows the oxygen dissociation pressure to form the oxide and the free energy for the elements forming oxides introduced above. It is clear that the formation of the oxide at the crack tip follows the thermodynamics suggesting the oxygen partial pressure gradient along the oxide tip ahead of the crack tip.

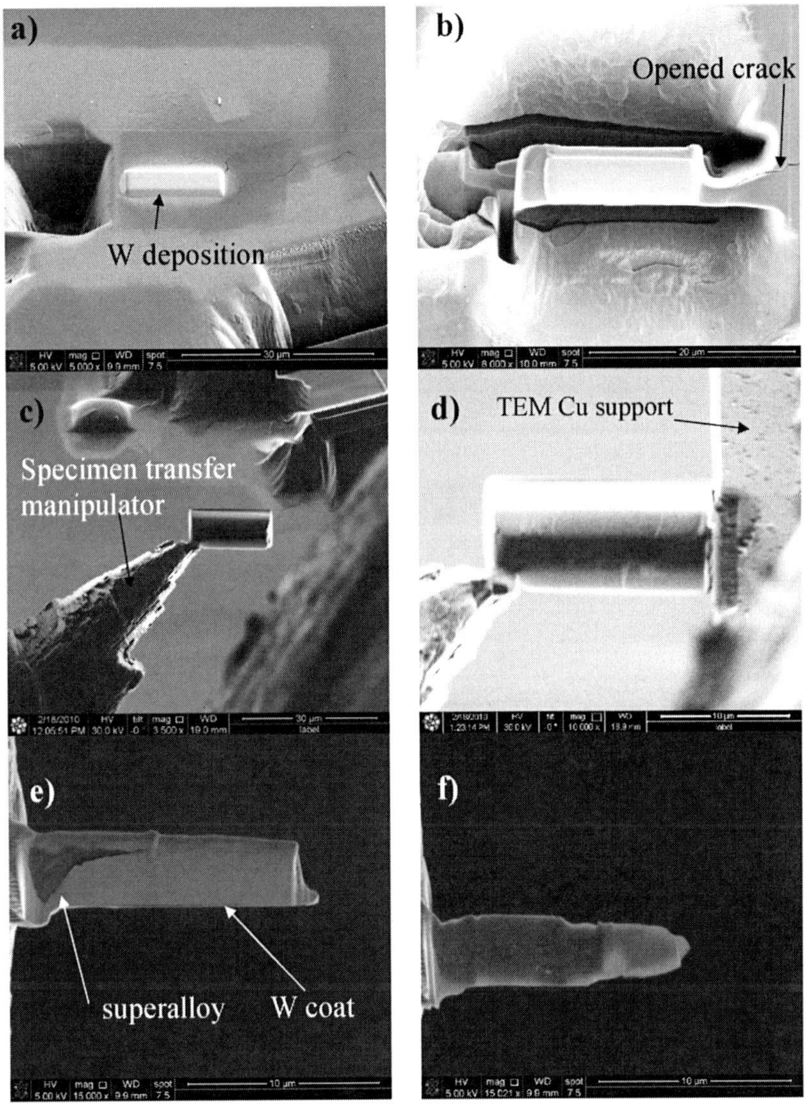


**Figure 23.** SEM secondary electron images of the process of the site specific TEM sample preparation

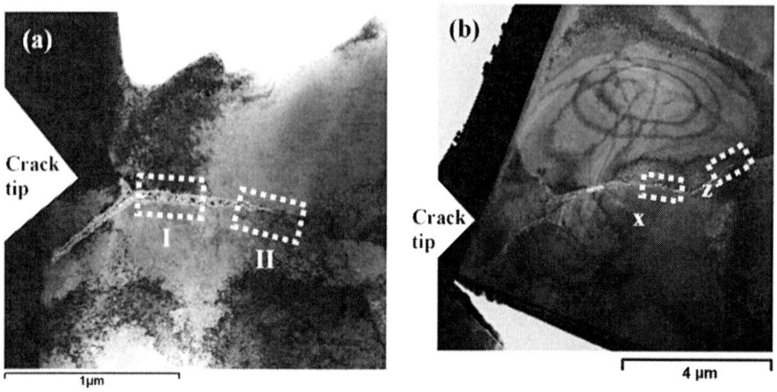


**Figure 24.** TEM bright field images of the two specimens. The crack tips locate just next to the specimen and the crack propagate from the left to right.

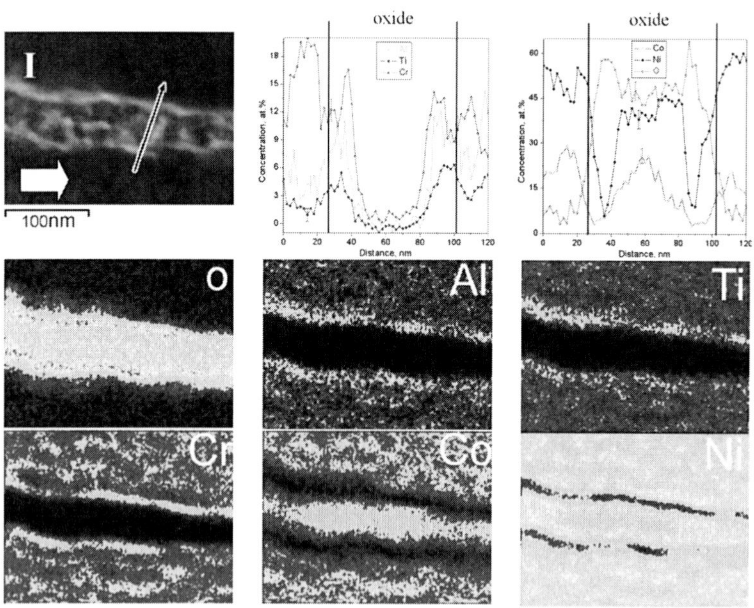


**Figure 25.** STEM dark field image from the region I in **Fig. 24** (top left); the thick white arrow indicates the crack growth direction and the thin black arrow indicates the area and the direction of the EDX line scan. The EDX line scan across the oxide (top right) and the EDX mapping results of the oxide (bottom).

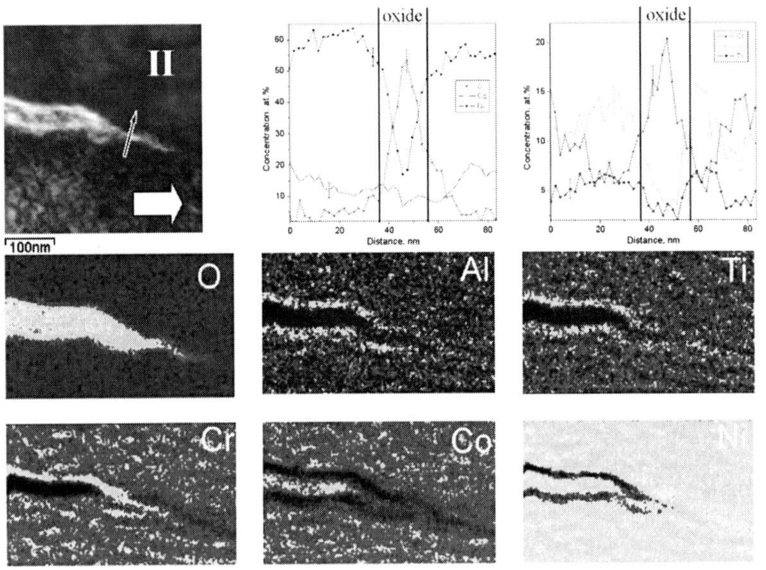


**Figure 26.** STEM dark field image from the region II in **Fig. 24** (top left). The EDX line scan across the oxide (top right) and the EDX mapping results of the oxide (bottom).

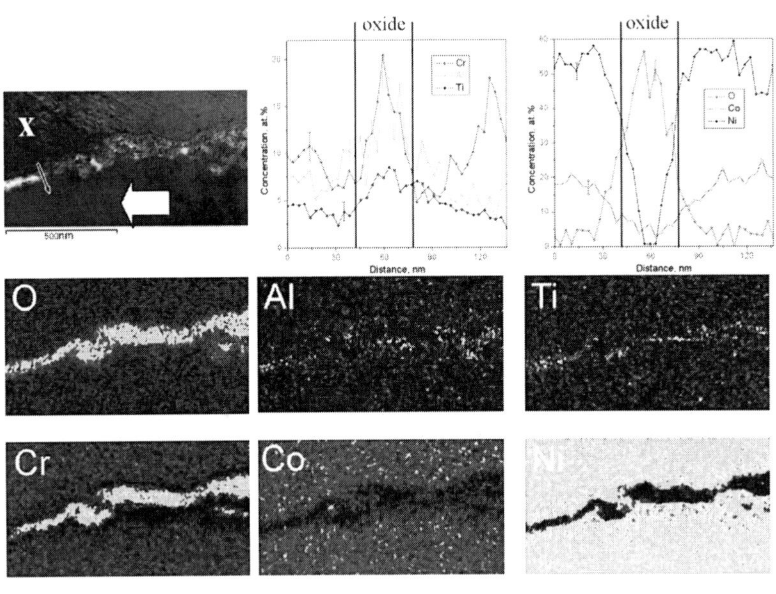


**Figure 27.** STEM dark field image from the region x in **Fig. 24** (top left). The EDX line scan across the oxide (top right) and the EDX mapping results of the oxide (bottom).

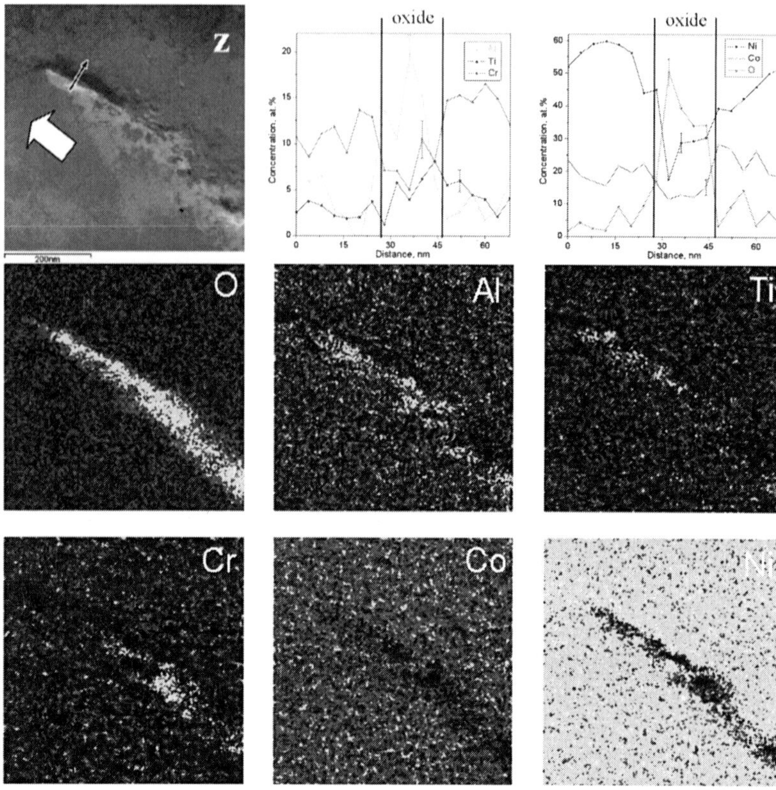


**Figure 28.** STEM dark field image from the region z in **Fig. 24** (top left). The EDX line scan across the oxide(top right) and the EDX mapping results of the oxide (bottom).

**Table 2.** Oxide formation free energy and the dissociation pressure at 727°C for the elements introduced in this section [**38**]

| | Oxide formation | Free energy [kJ/mole] | Oxygen dissociation pressure in $\log_{10}$ [bar] |
|---|---|---|---|
| Al | $Al_2O_3$ | -1425 | -48 |
| Ti | $Ti_2O_3$ | -1295 | -44 |
| | $TiO_2$ | -798 | -40 |
| Cr | $Cr_2O_3$ | -877 | -31 |
| Co | CoO | -166 | -18 |
| Ni | NiO | -149 | -16 |

# 6. SUMMARY

One of the challenges of the advanced Ni based superalloys is in the damage tolerance properties without reducing their superior strength at high temperature. The microstructure, particularly the grain boundaries, was found to be controlled by the two factors in this study. Firstly, it is the nominal chemical composition, especially the hardening precipitate participants. Secondly, heat treatment has a profound influence of microstructure.

The damage tolerance properties are also concerned with the environmentally assisted crack propagation along grain boundaries, which is essentially the oxidation assisted crack propagation in this study. In general, chromium oxide ($Cr_2O_3$) has been regarded as a healing agent of the oxidation process in Ni alloys. The state-of-the-art technique enabled us to observe the crack tip oxidation. In this study, it was successfully presented that the oxidation sequence is following the free energies for the oxides to form. Thus, for example, $Cr_2O_3$ is one of the earliest oxides to form at the crack tip.

Understanding the environmentally assisted crack propagation is one of the crucial fields of research to increase the turbine entry temperature (TET), which is indeed one of the most significant attainments in the propulsion industries.

# ACKNOWLEDGEMENT

The author would thank to Profs. I.P. Jones and P. Bowen at The University of Birmingham and Dr. M.C. Hardy at Rolls-Royce plc for their advice and the guidance. The joint support of the UK Engineering and Physical Sciences Research Council (EPSRC) and Rolls-Royce plc is also gratefully acknowledged. Many thanks are also due to Dr H.Y. Li, Research Fellow at the School of Metallurgy and Materials, The University of Birmingham and Dr Y.Y. Tse, former Research Fellow at the School of Metallurgy and Materials, The University of Birmingham, currently Lecturer at Loughborough University, for their assistance with the mechanical tests and the FIB TEM sample preparation.

# REFERENCES


1. G.W. Meetham, The Development of Gas Turbine Materials 1st ed., Applied Science, London, 1981.

2. R.C. Reed, The Superalloys, Cambridge University Press,Cambridge, 2006

3. H. Cohen, G. F. C. Rodgers, H. I. H. Saravanamuttoo, turbine. Gas, Third. theory, edition, 1987

4. C.T. Sims and W.C. Hagel, The Superalloys, John Wiley and Sons Inc., Canada, 1972.

5. L. R. Liu, T. Jin, N. R. Zhao, X. F. Sun, H. R. Guan, Z. Q. Hu, Science. Materials, A. Engineering, Materials. Structural, Microstructure. Properties, Processing, A3 2003 2003 191197 ,

6. X. Z. Qin, J. T. Guo, C. Yuan, J. S. Hou, H. Q. Ye, Letters. . Materials, 2008

7. S. T. Wlodek, M. Kelly, D. A. Alden, T. M. S. in, P. A. U. S. A. Warrendale, 1996129136

8. T. Sugui, W. Minggang, L. Tang, Q. Benjiang, X. Jun, Science. Materials, A. 5. Engineering, 2010

9. R. J. Mitchell, C. M. F. Rae, S. Tin, Science. Materials, . Technology, 2005

10. P.Villars, ASMInternational, 1997

11. K.R. Bain, M.L. Gambone, J.M. Hyzak, M.C. Thomas, Proceedings of the International Symposium on Superalloys,1988

12. D. Locq, P. Caron, S. Raujol, F. Pettinari-Sturmel, A. Coujou, N. Clement, Superalloys, 2004Proceedings of the Tenth International Symposium on Superalloys (2004) 179-187.

13. P. R. Bhowal, E. R. Wright, E. L. Raymond, Transactions. A. Metallurgical, Metallurgy. Physical, Science. Materials, A. , 1990

14. N. J. Hide, M. B. Henderson, P. A. S. Reed, S. U. P. E. R. A. L. L. O. Y. S. , 2000. Proceedings of the Ninth International Symposium on Superalloys, (2000) 495-503.

15. J. Telesman, P. Kantzos, J. Gayda, P. J. Bonacuse, A. Prescenzi, Superalloys, 2004Proceedings of the Tenth International Symposium on Superalloys (2004) 215-224.

16. J. Telesman, T. P. Gabb, A. Garg, P. Bonacuse, J. Gayda, Superalloys, 2008

17. E. Andrieu, R. Molins, H. Ghonem, A. Pineau, Science. Materials, A. Engineering, A1 1992 1992 2128 ,

18. H. Ghonem, D. Zheng, Science. Materials, A. Engineering, Materials. Structural, Microstructure. Properties, Processing, A1 1992 1992 151160 ,

19. M. Gao, D. J. Dwyer, R. P. Wei, Metallurgica. Scripta, Materialia. . et, 1995

20. A. K. Koul, R. Thamburaj, Transactions. A. Metallurgical, Metallurgy. Physical, Science. Materials, A. , 1985

21. D. Turan, D. Hunt, D. M. Knowles, Science. Materials, Technology, 2007

22. E. Andrieu, A. Pineau, Physique. I. V. Journal de, Proceedings, 1999

23. R. Molins, J. C. Chassaigne, E. Andrieu, Science. Materials, 2512. Forum, 1997

24. R. Molins, G. Hochstetter, J. C. Chassaigne, E. Andrieu, Materialia. . Acta, 1997

25. R. G. Baker, J. Nutting, Iron, Steel, 1959

26. G. Thomas, J. Nutting, of. Institute, Metals, Journal, 1957

27. The Journal of the institute of metals; Discussion, 911963

28. H. Gleiter, E. Hornbogen, science. Materials, engineering, 1968

29. E. Grant, A. Porter, B. Ralph, of. Journal, Science. . Materials, 1984

30. A. Porter, B. Ralph, of. Journal, Science. . Materials, 1981

31. R.D. Doherty, Metal Science,1619821982113

32. G. W. Lorimer, R. B. Hicholson, Metallurgica. . Acta, 1966

33. G.C. Wood, F.H. Stott, Materials Science and Technology, 319861986519530

34. L.L. Shreir and R.A. Jarman and G.T. Burstein, Corrosion, Butterworth-Heinemann Ltd,Oxford, 1963

35. S. Lozano-Perez, Y. Huang, R. Langford, J. M. Titchmarsh, Microscopy. Electron, Analysis, Proceedings, 5-7 Sept. 2001IOP Publishing, Bristol, UK, 2001, 191194

36. D. M. Longo, J. M. Howe, W. C. Johnson, Ultramicroscopy, 1999

37. F. A. Stevie, R. B. Irwin, T. L. Shofner, S. R. Brown, J. L. Drown, L. A. Giannuzzi, Characterization, for. U. L. S. I. Metrology, Technology, International Conference, 23-27 March 1998AIP, USA, 1998, 868872

38. G.V. Samsonov, The oxide handbook, Plenim Publishing Company, Ltd, London,197323

4

# GOLD NANOSTRUCTURES PREPARED ON SOLID SURFACE

**Jakub Siegel[1], Ondřej Kvítek[1], Petr Slepička[1], Václav Švorčík[1] and Zdeňka Kolská[2]**

[1] Institute of Chemical Technology Prague, Czech Republic

[2] University of J.E. Purkyne Usti nad Labem, Czech Republic

## 1. INTRODUCTION

Up to now, many efforts have been made to produce smart materials with extraordinary properties usable in broad range of technological applications. In particular, within the last two decades, it has been demonstrated that properties of new prospective materials depend not only on their chemical composition but also on the dimensions of their building blocks which may consist of common materials [1,2].

A nanoparticle consists of a few atoms forming a cluster with size in the nanometer range. A nanometer represents a magical size of matter around which the vast majority of materials possess extraordinary, novel physico-chemical properties compared to its bulk form. Considerable attention has been focused during the last few decades on developing and optimizing methods for the preparation of gold nanoparticles to size and shape. Especially properties of 0D spherical and non-spherical particles, as applications of nanostructured materials, may differ considerably depending on the particle shape itself. Simple and straightforward example of the shape dependent behaviour of nanometer-sized particles is its colour. Ultrasmall gold spheres or clusters has been known for centuries as the deep red ruby colour of stained glass windows in cathedrals and domestic glassware. The colour results from the plasmon resonances in the metal cluster. Nowadays, most gold nanoparticles are produced via wet, chemical routes. Nevertheless, synthesis of metal nanoparticles (NPs) has been extensively studied since early 80's [3-9]. Some pioneering works on synthesis of gold nanoparticles were even published as far back as in early 50's by Turkevich [6]. Since that, many techniques have been developed, however, predominately based on wet, chemical processes [4-9]. Currently the most common noble-metal nanoparticle synthesis techniques are those developed by Brust-Shiffrin in 1994 [5]. The method based on reduction of AuIII+ complex compound with NaBH4 stabilized by thiols enables preparation of high stable particles with pretty narrow distribution and well-controlled size around 1 nm.

Besides interesting properties of nanostructured gold systems such as catalytic effects or magnetism [2,10], which both originate from surface and quantum size effects, they are also extremely usable those, which are closely connected with the average number of atoms in the nanoparticles. The properties and behavior of extremely small gold particles completely differ from those of bulk materials, e.g., their melting point [2,11,12], density [13], lattice parameter [13-15], and electrical or optical properties [13,14,16] are dramatically changed. Gold is also critical component in certain therapies, more specifically, in the treatment of cancer by hyperthermia and thermoablation. These two therapies use heat to kill cancer cells. In the case of hyperthermia, the cancerous tissue is heated to enhance conventional radiation and chemotherapy treatments, while in thermoablation the tissue is heated so that the cancer tissue is destroyed by the localised heat. In principle, there are two methods which can be used to provide heating, i.e. infra-red absorption and the application of an oscillating magnetic field to magnetic nanoparticles. Owing to this, nanosized gold is nowadays used in a vast range of cancer therapy applications such as cancer therapy agents [17] or cancer cell imaging [18,19]. Moreover, gold nanoparticles have often been conjugated with antibodies [20], or grafted to other carriers for surface property enhancement [21,22].

Besides above mentioned 0D nanostructures (nanodiscs, nanoparticles, nanoclusters) increasing efforts have been recently devoted also to one-dimensional (1D) nanostructures. 1D nanostructures in the form of wires, rods, belts and tubes have long been the focus of intensive research owing to their unique applications in mesoscopic physics and fabrication of nanoscale devices [23-25]. It is generally accepted that 1D nanostructures provide a good system for the investigation of the dependence of electrical and thermal transport or mechanical properties on dimensionality and size reduction (quantum confinement) [26]. Of the many elements and compounds from which nanowires may be made, gold is technologically important for its low electrical resistivity (2.21 μΩ cm) [27], its inertness to attack by air and its resistance to sulfur-based tarnishing [28]. Additionally, gold is more biocompatible than most metals, rendering it suitable for implantation [29,30] or electrical interfacing with cells [31,32] and tissues in nanobiological applications [33-35].

Nanostructured materials with high aspect ratios such as nanorods, nanowires, and nanoline patterns often exhibit anisotropic electronic and optical properties that differ from those observed in the bulk materials. These unique materials can be used to create many interesting devices, including fast responding chemical and biochemical sensors [36-40]. The high aspect ratio of nanowires should also make them interesting for the use in two dimensional photonic crystals, where vertical nanowires would constitute an array of high refractive index pillars in air [41]. Field emission from nanowires has also been reported [42], suggesting the possibility of devices such as field emission displays (FEDs) with nanowires acting as cathodes.

A variety of fabrication techniques have been developed in the past decade that yield high quality nanowires. Fabrication of ordered arrays of metallic nanoparticles supported on transparent substrate by sequential techniques like electron beam lithography has been demonstrated [43]. Such top-down approaches, however, are cumbersome and have a low yield, which hinders practical applications. High throughput approaches for the synthesis of metallic nanowires are thus intensely searched [44-48]. In general, the production of arrays of nanostructures on substrates by lithographic techniques presents the disadvantage of high cost and a restriction in the number of materials to which it can be applied. The method can also prove to be complex and inefficient. Template based methods overcome those disadvantages, but the obtained structures often present a high number of imperfections due to packing defects in the original templates [49].

Above mentioned applications, however, usually require gold nanostructures (0D or 1D) to be either suspended in colloid solution or attached to another support medium. Concerning this, creation of nanostructured gold directly on appropriate support may be technologically valuable since one can avoid additional preparation step oriented on metal-substrate mutual attachment. Therefore, this chapter focuses on new possible approaches for nanostructuring of gold layers either formerly deposited on solid substrates (polymer or glass) or during deposition itself (polymer). The formerly mentioned technique is based on the intensive post-deposition thermal annealing of sputtered layers on polytertaflouroethylene (PTFE) or glass, whereas the latter technique is based on forced (preferential) growth of gold on nanostructured polymer template. The method, combining nanoscale patterning of the polyethyleneterephtalate (PET) substrate by polarized light of excimer laser with glancing angle deposition of the gold, provides an interesting alternative to time consuming sequential lithography-based nanopatterning approaches.

First part of the text is focused on the study of selected physico-chemical properties of deposited gold layers and its changes induced by post-deposition annealing. The gold nanostructures of different thicknesses were sputtered onto glass or polymer (PTFE) substrate and then the samples were annealed from room temperature to 300°C. The effects of annealing on gold structures sputtered onto substrate, their surface morphology and roughness were studied using Atomic Force Microscopy (AFM), lattice parameter and crystallites size and their distribution by X-ray diffraction (XRD) and by SAXSess. Hall mobility, volume resistance and free carrier concentration were measured by Van der Pauw method, an electric permitivity by ellipsometry, an optical band gap by UV-Vis spectroscopy and a sheet resistance of gold nanostructures by 2-point method.

In the next part special attention will be given to the irradiation of PET surface with linearly polarized light of a pulsed KrF excimer laser to produce templates for preparation of laterally ordered self-organized arrays of metallic nanowires. Different fluences and angles of incidence of the laser beam were applied. The periodicity of the ripples created on the polymer surface was controlled by changing the incidence angle of laser light during irradiation. Subsequently the modified polymer surface was coated with gold using two deposition techniques (sputtering and evaporation). The surface of nano-patterned coated/uncoated PET was analyzed by AFM and a scanning electron microscopy equipped with a focused ion beam (FIB-SEM), allowing to cut cross-sections of the laser patterned substrate surface and the deposited gold layers

# 2. GOLD NANOSTRUCTURES ON GLASS SUBSTRATE

An overview of growth process, morphology, electrical and optical properties of ultra-thin gold layers sputtered on glass is provided in following sections. Insight into the phenomena taking place during post-deposition thermal treatment is also given.

## 2.1. Thickness, Morphology and Inner Structure

Thickness of sputtered layers was measured by AFM. Thickness in the initial stage of deposition (sputtering time less than 50 s) was determined from the SEM image of the sample cross-section (FIB-SEM). Dependence of the layer thickness on sputtering time is shown in Fig. 1. Linear dependence between sputtering time and structure thickness is evident even in the initial stage of the layer growth. This finding is in contradiction with results obtained earlier for Au sputtering on PET [50]. In that case, the initial stage of the layer growth was related to lower deposition rate which is due to different morphology.

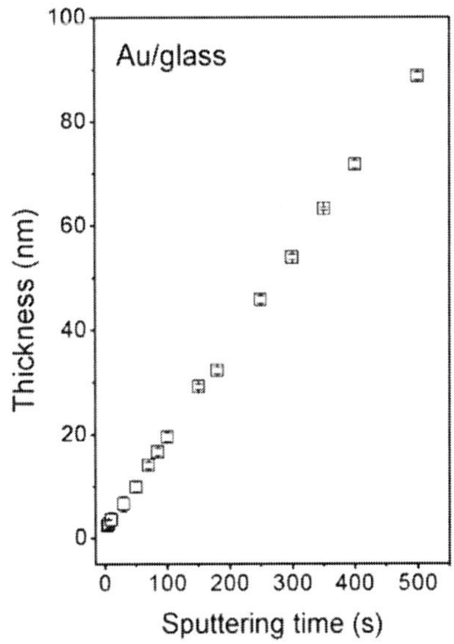


**Figure 1.** Dependence of the gold structure thickness on sputtering time [13].

In Fig. 2, a SEM picture of the cross-section of the Au layer at its initial stage of growth is shown. It is obvious that after approximately 20 s of

Au deposition, flat, discrete Au islands (clusters) appear on the substrate surface. The flatness may indicate preferential growth of gold clusters in a lateral direction. When the surface coverage increases and the clusters get in close contact with each other, a coarsening sets in and becomes the dominant process. After the surface is fully covered, additional adsorption causes only the vertical layer growth, while the lateral growth is dominated by cluster boundary motion [51].

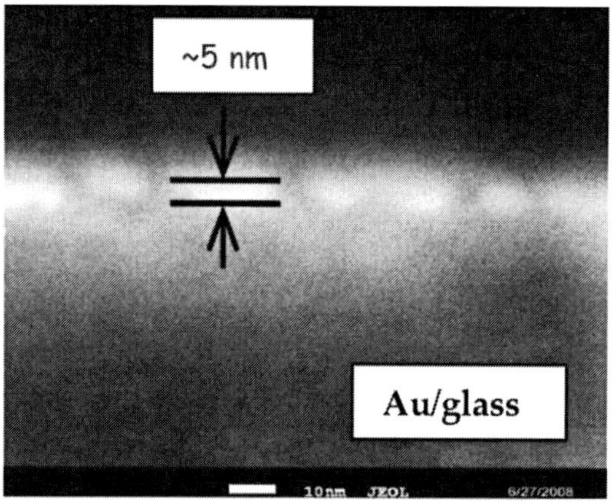


**Figure 2.** SEM scan of the FIB section of gold structure on glass substrate. Deposition time was 20 s. [13].

The AFM images that illustrate the surface morphology and roughness (Ra) of gold-coated glass before and after annealing are shown in Fig. 3. For the sake of comparison only images of the samples with identical vertical scale were chosen. From Fig. 3 it is clear that the surface morphology of the as-sputtered structures does not depend significantly on the sputtering times. Monotonous decrease of surface roughness with deposition time is related to the stage of the layer growth. During initial stages of metal growth the layer is formed over isolated islands. After that, during ongoing deposition, interconnections between clusters are formed and the deposited layers become homogeneous and uniform. Decrease of surface roughness is direct evidence of the formation of a thicker layer during sputtering process on flat substrate. After annealing, however, the surface morphology changes dramatically. Similar changes in the morphology of the thin gold structures have also been observed on the samples annealed at 200°C for 20 hours [52] and at 450°C for 2 hours [11]. It is seen from Fig. 3 that the annealing leads to the formation of "spherolytic and hummock-like" structures in the gold layers. The formation may be connected with an enhanced diffusion of gold particles at elevated temperature and their aggregation into larger structures. It is well known that the melting point of the gold nanoparticles decreases rapidly with decreasing particle size [2,11,12].

The migration of the gold nanoparticles and formation of larger structures may be connected with lower thermodynamic stability of the gold nanoparticles and lower gold wettability of glass. This idea is supported by some previous XRD experiments in which dominant (111) orientation of gold crystals in the sputtered gold layers was determined [11, 53]. The (111) oriented gold crystals are known to be thermodynamically unstable and their melting and cracking starts from the edge parts that should be bounded to Au (110) surface [11].

Metallic nanoparticles and generally nanostructures composed of metals often exhibit different values of structure parameters compared to their bulk form e.g. contraction of lattice parameter in nanostructures increases material density [13,53]. Lattice parameters a of the face-centered cubic gold nanostructures determined before and after annealing are shown in Fig. 4 as a function of the sputtering time (i.e. effective layer thickness). Lattice parameters were calculated using the Rietveld procedure (full pattern fitting). For this purpose the five strongest diffraction lines were taken into account. For very thin films the diffraction lines are weak and the resulting values of the lattice parameters are loaded by a higher error. The error is especially large for the as-sputtered samples. A dramatic difference is found in the dependences of the lattice parameter on the sputtering time between as-sputtered and annealed samples. For as-sputtered samples the lattice parameter varies rapidly with the increasing sputtering time, i.e. with the increasing mean size of the gold crystallites [16,53,54]. It is seen that a maximum lattice parameter is observed after 100 s of sputtering, i.e. for the layer thickness of about 18 nm. For both the thinner and thicker layers the lattice parameter declines significantly. The same trend in the lattice parameter vs. sputtering time dependence was reported also for silver structures, for which the lattice parameter increases slightly up to the structures size of 12 nm and then decreases [55]. In contrast to the as-sputtered gold structures at the annealed ones the lattice parameter is nearly independent on the sputtering time and the size of the structures.

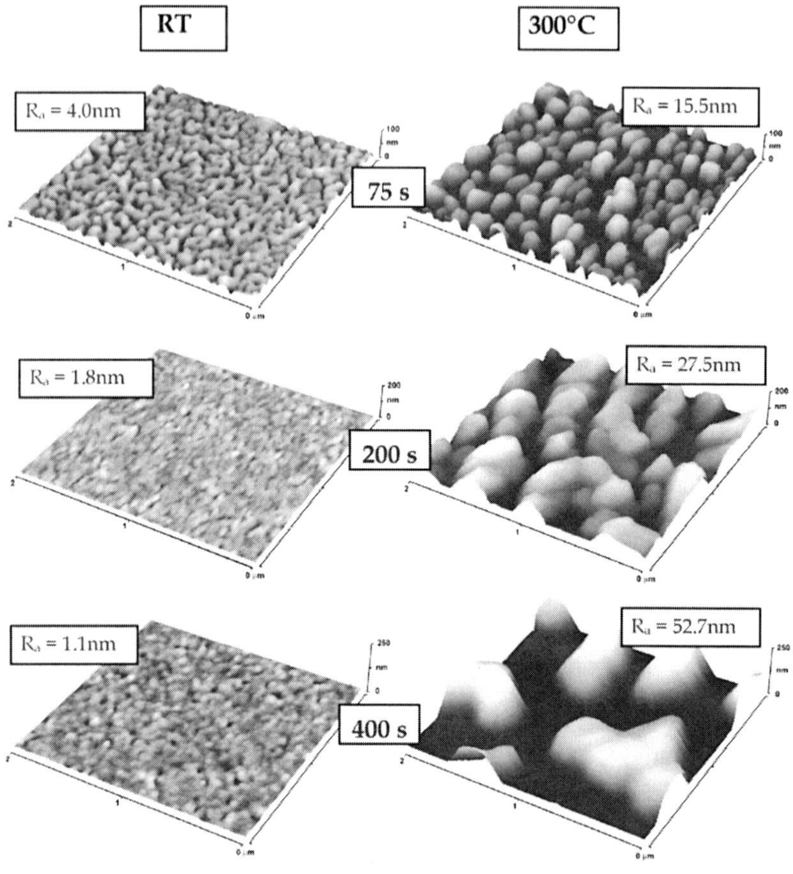


**Figure 3.** AFM scans of gold structures sputtered for 75, 200 and 400 s on glass substrate before (RT) and after annealing (300°C). Ra is the average surface roughness in nm [16].

The dependence of the crystallite size on the sputtering time before and after annealing is shown in Fig. 5. The dependences are quite different for the as-sputtered and annealed samples. While in the as-sputtered samples the crystallite size is amonotonously increasing function of the sputtering time, in the annealed ones the crystallite size first increases rapidly up to the sputtering time of 250 s, achieves a maximum and then it decreases. A dramatic increase of the Au crystallite size with the annealing temperature was published also by Santos et al. [56].

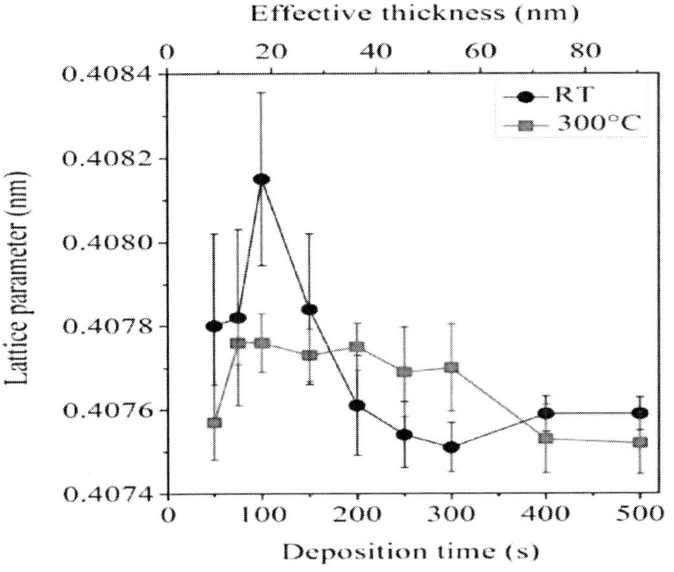


**Figure 4.** Dependence of the gold lattice parameter on the sputtering time (i.e. effective thickness) measured before (-•-) and after annealing at 300°C (-■-)[14].

Size distribution of the Au crystallites determined by SAXSess method is presented in Fig. 6. The mean crystallite size values (modus) determined by SAXSess (S) and by XRD (X) are compared in Fig. 5. It is obvious that before annealing both methods (SAXSess, XRD) give the same values of the mean crystallite size, which increase slightly with the deposition time. Both SAXSess and XRD measurements prove dramatic increase of the mean crystallite size after annealing. However, there is an obvious dissimilarity between SAXSess and XRD results regarding longer sputtering time which is caused by inability of the SAXSess method to examine crystallites larger than ca 90–100 nm. The different behavior may be due to a crystallites' re-crystallization in the annealing process. The crystallite size determined by the XRD technique is based on the determination of the so-called coherently diffracting domains with their mean dimensions in direction perpendicular to the film surface. This is the reason why the crystallites determined in this way significantly exceed their size in some cases.

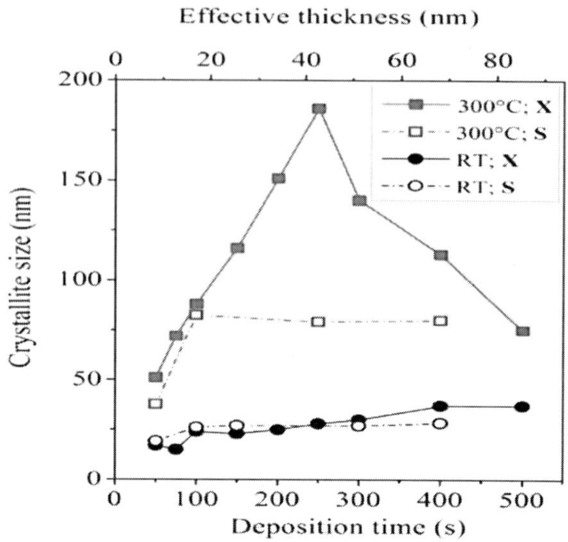


**Figure 5.** Dependence of the size of the gold crystallites on the sputtering time (i.e. layer effective thickness) measured before (-○- S, -●- X) and after annealing at 300°C (-□- S, -■- X) using S – SAXSess, X – XRD methods [14].

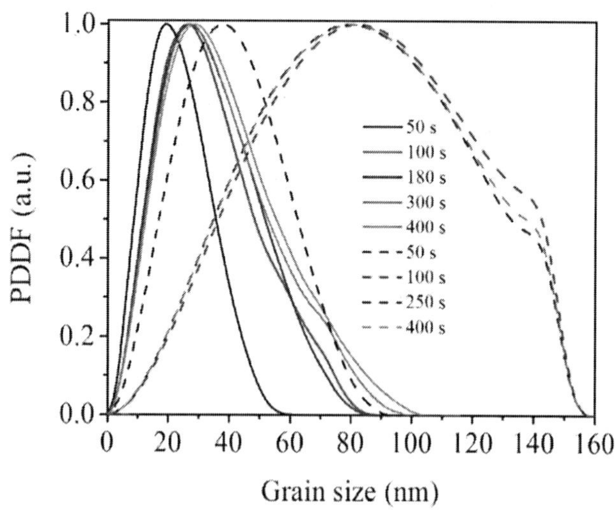


**Figure 6.** Pair distance distribution functions (PDDF) of gold crystallites by different sputtering time (in seconds) before (solid line) and after annealing (dashed line) measured by SAXSess method [14].

## 2.2. Optical and Electrical Properties

Besides interesting catalytic and electronic properties, nanoparticles of noble metals exhibit also distinctive, shape-dependent optical properties that have attracted great technological interest. This is particularly true for gold nanostructures [11]. Images of the sample surface for different sputtering times and for sputtered and annealed samples are shown in Fig. 7.

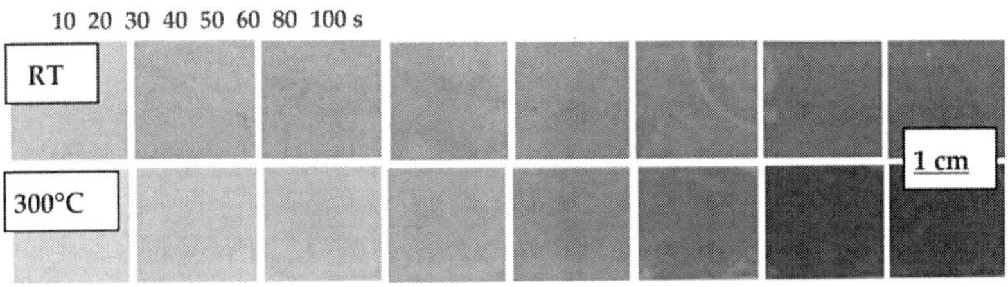


**Figure 7.** Images of the glass samples with gold structures sputtered for increasing times. The as-sputtered (RT) and annealed samples (300°C) are shown for comparison [16].

The deposited samples become darker with increasing sputtering time, the darkening being related to increasing thickness of the gold structures. Also a gradual change of the structures colour from blue to green is seen. After annealing all structures exhibit reddish colour, regardless of the sputtering time. The changes in the layer colour indicate pronounced alteration in the gold nanostructure caused by the annealing (see Fig. 3). It could be in accordance with previously presented results, the small gold sample about 10 nm absorbs green light and thus appears red [2]. This effect was confirmed also by UV-Vis spectroscopy. For the sake of clarity only some of UV-Vis spectra from as-sputtered and annealed samples are shown in Fig. 8 (RT and 300°C). The absorbance of gold structures increase with increasing sputtering time and structure thickness as could be expected. From comparison of the spectra of the sputtered and annealed samples it is seen that the annealed structures have qualitatively different shapes and lower absorbance. Both phenomena point at structural changes due to annealing. The observed shift of the 530 nm absorption peak (corresponding to surface plasmon resonance) with increasing sputtering time towards longer wavelengths is probably related to interconnection and mutual interaction of gold nanosized islands in the structure. From present UV-Vis spectra it is evident that the as-sputtered samples prepared for deposition time of 30 s and that annealed one (sputtering time 200 s) are the first ones, which do not exhibit the peak of plasmon resonance. Qualitative difference between absorbances of the sputtered structure and that annealed may indicate a transition from the structure comprising discrete gold islands to continuous gold coverage.

The UV-Vis spectra were also interpreted in the frame of the well-known Tauc's model [57] and the optical band gap (Egopt) was calculated as a function of the sputtering time for sputtered and annealed samples. This dependence is shown in Fig. 9. Also the gold structure thickness vs. sputtering time measured by AFM method on the scratch step is presented. These values, inclusive the result that the AFM-scratch technique is not applicable on annealed structures due to their altered morphology, were taken from our previous work [53]. It is seen from Fig. 9 that the structures sputtered for times below 150 s exhibit non-zero Egopt. Rather dramatic change in the Egopt is observed after annealing, where the values of the Egopt are much higher in comparison with those of the sputtered sample. For samples sputtered for times around 300 s a non-zero Egopt is observed. For behaviour of ultra-thin metal structures (<10 nm) the surface-size and quantum-size effects must be considered [2,53,58]. This quantum-size effect in small structures leads e.g. to a semi-conducting character, which is accompanied by non-zero Eg (band gap) or Egopt. This effect was observed in the present case.

The dependence of the volume resistivity on the sputtering time is seen from Fig. 10a. For as-sputtered samples a rapid drop of the resistivity over a narrow thickness interval is observed. The drop indicates a transition from electrically discontinuous to continuous gold coverage. For annealed samples the resistivity drop is shifted towards thicker layers. The difference is obviously connected with changes in the layer structure taking place during annealing i.e. gold coalescence and formation of isolated islands [16]. The onset of the rapid resistivity drop is observed after 50 and 150 s of sputtering for as-sputtered and annealed samples, respectively. Free carrier volume concentration and their Hall mobility significantly affect the electrical conductance of materials. The dependence of the free carrier concentration and the mobility on the sputtering time is shown in Figs. 10b and 10c, respectively. As can be seen from Fig. 10b, with increasing sputtering time the carrier concentration increases dramatically and the layers become conductive (see Fig. 10a). As in the case of resistivity the onset of the rapid increase of the free carrier concentration on annealed samples is shifted towards longer sputtering time. Thus the constant level of the free carrier concentration is achieved later compared to the as-sputtered samples. A similar dependence of the free carrier concentration on the layer thickness was recently observed on PET and PTFE sputtered with gold [59]. The carrier mobility also changes dramatically with increasing sputtering time for non-annealed and annealed samples (Fig. 10c). The mobility first declines rapidly to a point when an electrically continuous layer is formed. The decline may be due to the fact that in a discontinuous layer the mobility mechanism differs from classical electron conductivity common in metals. For annealed structures the continuous layer is formed after a longer deposition time. For thicker, electrically continuous gold layers the mobility is a slowly increasing function of the sputtering time.

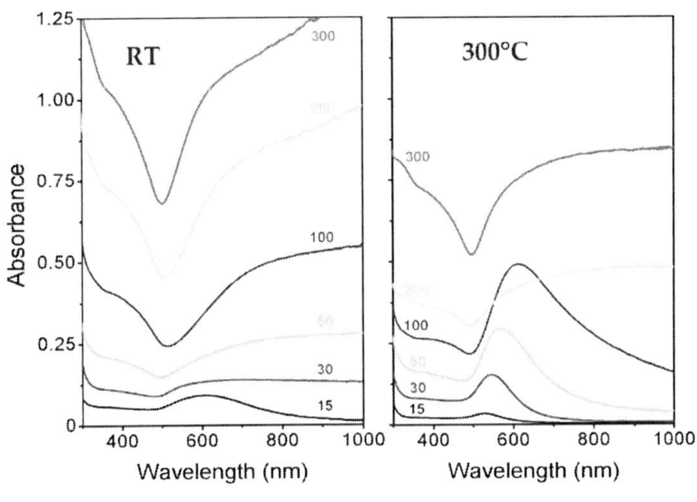


**Figure 8.** UV-Vis spectra of gold structures sputtered on glass before (RT) and after annealing (300°C). The numbers in Figs. are sputtering times in s [16].

There is a clear correspondence between mobility (Fig. 10c), free carrier volume concentration (Fig. 10b) and volume resistivity (Fig. 10a). A similar dependence of the free carrier concentration on the thickness of the gold layers deposited by sputtering on PET and PTFE was observed [59]. Simple and straightforward interpretation of the above described observations is that during electrical measurement on discontinuous gold layers an electron injection due to the tunneling effect occurs [13,16]. With ongoing deposition time the discrete structures become interconnected and form an electrically continuous, homogeneous layer in which the concentration of free carriers is saturated.

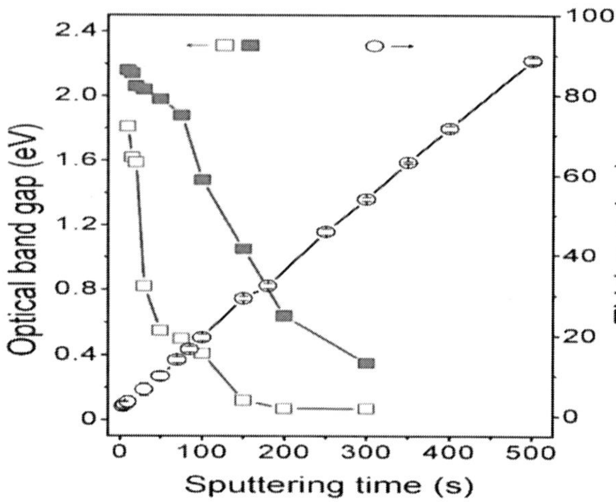


**Figure 9.** Dependence on the sputtering time of the optical band gap of gold structures before (RT) and after annealing (300°C) (-□- for RT and -■- for 300°C) and thickness (-○-) [16].

The IR part of optical constants of the as-deposited and annealed Au films determined from ellipsometry also supports the results of electrical transport measurements. Fig. 11 presents the real part of electric permittivity in the studied spectral range.

Spectroscopic features in the Drude (IR) region clearly show the tendency of Au films to lose their metallic behavior with decreasing thickness due to gold coalescence, leading to a layer discontinuity [16]. Film discontinuity of the as-deposited thin layers is a natural consequence of the mechanism of the layer growth. Percolation threshold is reached at the layer thickness of about 7 nm [60] corresponding to the deposition time of about 25 s in the present case. Fig. 11 also shows that the strong change in the surface morphology induced by the annealing shifts the metal-to-insulator transition towards greater layer thicknesses (i.e. deposition times). The thickness variation of IR end of the real part of the electric permittivity spectra of annealed gold layers (positive value reaching the maximum and then passing through zero to negative values with increasing deposition times) is consistent with previous studies of metallic films around the percolation threshold [60]. For the annealed layers with sputtering times equal to or smaller than those corresponding to the metal-to-insulator transition, a strong signature of plasmons is expected in the VIS part of optical constants. This is documented in Fig. 12, where the spectral dependence of the imaginary part of the electric permittivity is shown.

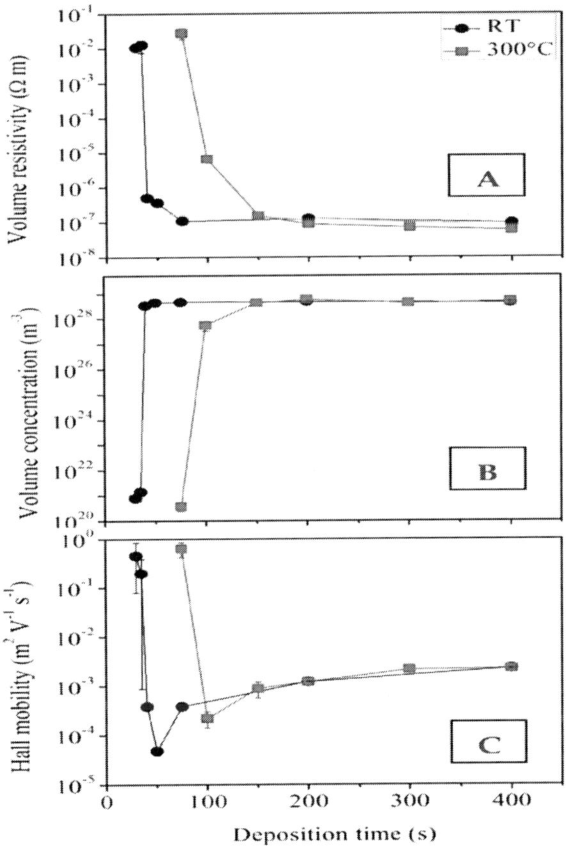


**Figure 10.** Dependence of the volume resistivity (A), surface free carrier volume concentration (B) and surface free carrier Hall mobility (C) on the sputtering time measured by van der Pauw technique before (-●-) and after annealing at 300°C (-■-) [14].

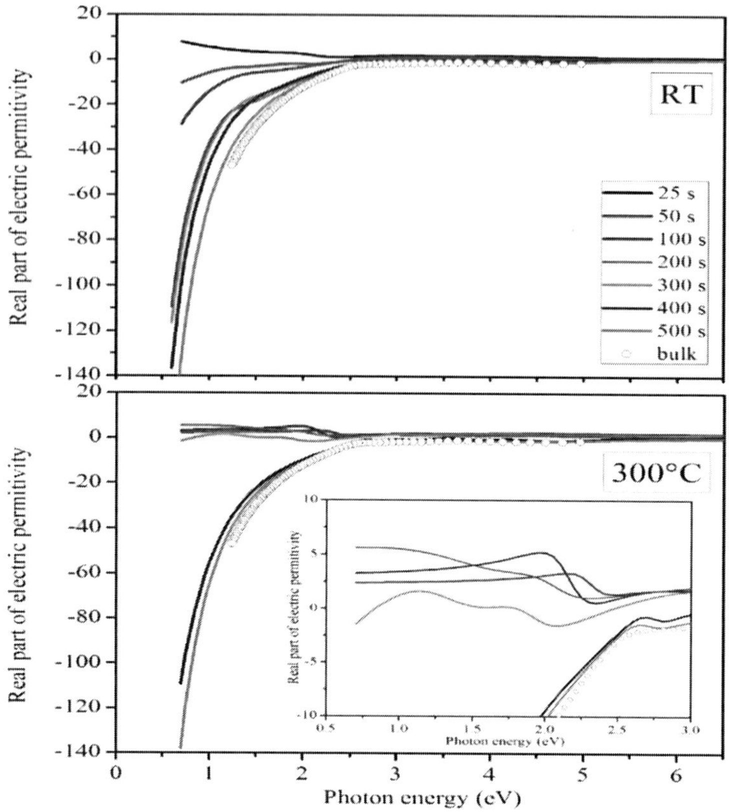


**Figure 11.** Real part of the electric permittivity spectra of as-sputtered (RT, upper part) and annealed (300°C, lower part) gold structures obtained by spectroscopic ellipsometry for the different sputtering times [14].

Plasmon oscillator band for the layer sputtered for 50 s is centered at around 2.3 eV (540 nm). With increasing deposition time the band becomes broader and shifts to longer wavelengths (lower photon energy). For 200 s sputtering time the plasmon band splits into two and for longer sputtering times it integrates into the Drude term in the IR spectral limit. This change in the optical constants around the metal-to-insulator transition is the reason for the color variation of the annealed layers.

The temperature dependence of the sheet resistance for two particular structure thicknesses is displayed in Fig. 13. One can see that the temperature dependence of the sheet resistance strongly depends on the structure thickness. For the layer about 89 nm thick, the resistance is an increasing function of the sample temperature, the expected behavior for metals. For the structure about 6 nm thick, the sheet resistance first decreases rapidly with increasing temperature, but above a temperature of about 250 K, a slight increase in resistance is observed. The initial decrease and the final increase of the sheet resistance with increasing temperature are typical of semiconductors and metals, respectively. It has been referred elsewhere [2] that a small metal cluster can exhibit both metal and semiconductor characteristics just by varying the temperature. It is due to temperature-affected evolution of band gap and density of electron states in the systems containing low number of atoms.

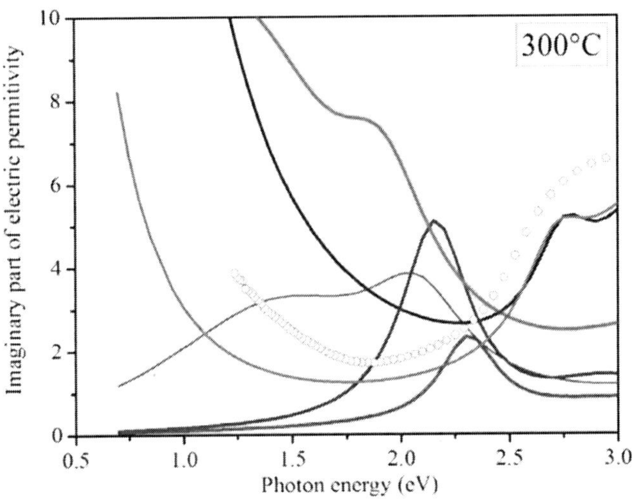


**Figure 12.** Imaginary part of the electric permitivity spectra of annealed (300°C) gold structures. The presence and evolution of the plasmon bands should be noted (for details see the text). Coloured sign of curves is the same as in Fig. 11 [14].

From the present experimental data, it may be concluded that for the thicknesses above 10 nm, the sputtered gold layers exhibit metal conductivity. In the thickness range from 5 to 10 nm, the semiconductor-like and metal conductivities are observed at low and high temperatures, respectively. Our further measurements showed that the layers thinner than 5 nm exhibit a semiconductive-like characteristic in the whole investigated temperature scale. Except for band gap evolution theory, typical semiconductor-like behavior may also originate from the tunneling effect of electrons through the discontinuous, separated Au clusters during electrical measurements. Since the probability of electron tunneling depends on the temperature, similarly, typical course of sheet resistance and, as will be shown later, CV characteristic may be affected right by this phenomenon.

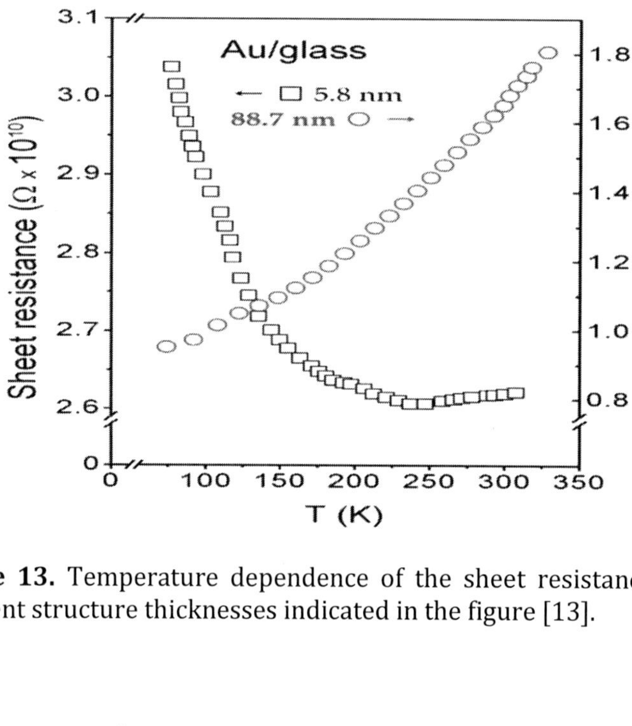


**Figure 13.** Temperature dependence of the sheet resistance for two different structure thicknesses indicated in the figure [13].

From presented measurements of sheet resistance results the semiconductor-like character of Au at specific structure conditions (thickness, temperature). The observed semiconductor-like character (decreasing resistance with increasing temperature) of ultrathin Au structures may originate from two undistinguishable phenomena. The first one results from a tunneling effect which occurs at discontinuous structures during resistance measurements [59]. The second one originates from the semiconductor characteristic of the intrinsic cluster itself, which occurs in metal nanostructures of sufficiently small proportions [2].

# 3. GOLD NANOSTRUCTURES ON POLYMERIC SUBSTRATE

In this section special attention is given to the changes in surface morphology and other physico-chemical properties of gold nanolayers, sputtered on polytetrafluoroethylene (PTFE) surface induced by post-deposition annealing.

## 3.1. Electrical Properties

The dependence of the electrical sheet resistance (Rs) of the gold layer on its thickness before and after annealing (at 100, 200 and 300°C) is shown in Fig. 14. For the as-sputtered samples the sheet resistance decreases rapidly in the narrow thickness range from 10 to 15 nm when an electrically continuous gold coverage is formed. The resulting sheet resistance is saturated at a level of aproximately 200 Ω. From the measured Rs and effective layer thickness, layer resistivity R (Ohm centimeter) was calculated, which appears to be few orders of magnitude higher than that reported for metallic bulk gold (RAubulk = 2.5 × 10−6 Ω cm [61], e.g., for 100 nm thick Au layer RAu100 nm = 1 × 10−3 Ω cm). As in the case of Au coated glass substrate (see section 2), the higher resistivity of thin gold structures is due to the size effect in accord with the Matthiessen rule [62]. Annealing at temperatures below 200°C causes only mild shift in the resistance curve towards thicker layers. Transition from electrically discontinuous to electrically continuous layer in case of low temperature annealed samples is more gradual and occurs between the effective layer thicknesses from 10 to 20 nm regarding the annealing temperature. After annealing at 300°C a dramatic change in the resistance curve is observed. The annealed layers are electrically discontinuous up to the Au effective thickness of 70 nm above which the continuous coverage is created and a percolation limit is overcome. However, for longer sputtering times up to 550 s, the sheet resistance changes slowly and it achieves a saturation which is observed on the as-sputtered layers and layers annealed at low temperatures.

Compared to electrical properties discussed in chapter 2 (Au layers on glass substrate), one can see that in case of PTFE substrate the transition from electrically discontinuous to continuous layer is shifted towards thicker layers. This fact is due to incomparable value of surface roughness of substrate used which is in the case of PTFE one order of magnitude higher (see section 3.3).

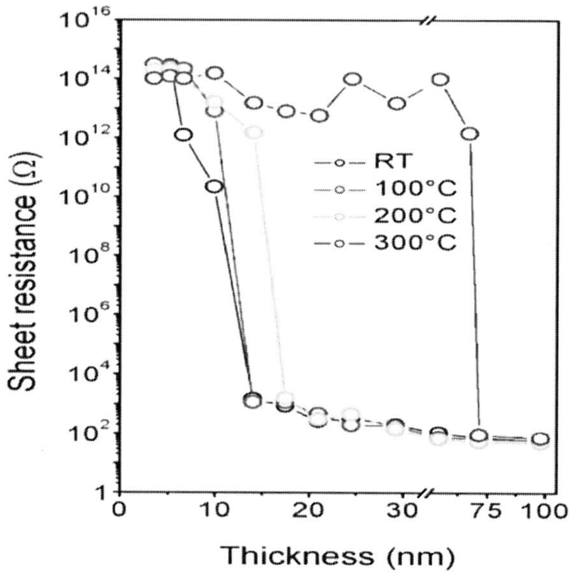


**Figure 14.** Dependence of the sheet resistance (Rs) on Au layer thickness for as-sputtered samples (RT) and the samples annealed at 100, 200 and 300°C [63].

## 3.2. Chemical Composition

Besides the sheet resistance measurements, information on the layer structure and homogeneity can be obtained in another way too. Here, complementary information on the layer homogeneity is obtained from XPS spectra. Fig. 15 A,B shows intensity normalized XPS spectra (line Au 4f) of 20 and 80 nm thick sputtered gold layers, respectively. Black line refers to as-sputtered layer and blue line to the one annealed at 300°C. Annealing of the 80 nm thick gold layer does not change the XPS spectrum. In contrast, the annealing of the 20 nm thick layer results in strong broadening of both lines which is due to the sample charging in the course of the XPS analysis. The charging is closely related to the change in the layer morphology: from electrically continuous one for as-sputtered sample to discontinuous one after the annealing procedure

[16]. This observation is in agreement with above described results of the sheet resistance measurements (see Fig. 1, section 3.1).

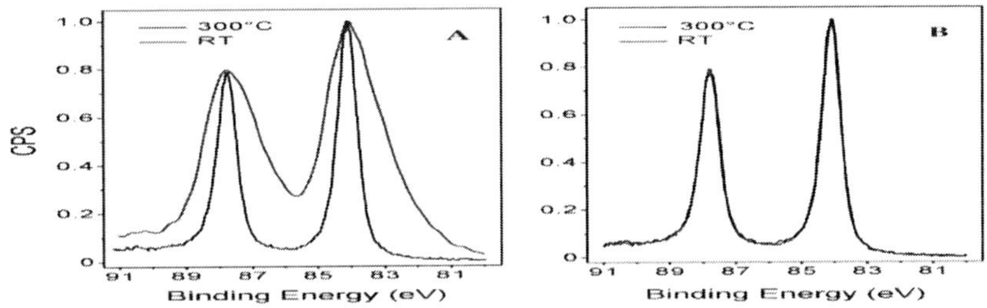


**Figure 15.** Intensity normalized XPS spectra (line Au (4f)) of 20 (A) and 80 nm (B) thick sputtered Au layers on PTFE before (black line) and after (blue line) annealing at 300°C [63].

Concentrations of chemical elements on the very sample surface (accessible depth of 6 to 8 atomic layers) determined from XPS spectra are summarized in Table 1. The XPS data were obtained for the samples with 20 and 80 nm thick gold layers, both as-sputtered and annealed at 300°C. Total carbon concentration and the carbon concentration coming from PTFE (calculated from XPS data) are shown in columns 1 and 2 of the table, respectively. Major part of the carbon is due to sample contamination. Fluorine to PTFE carbon ratio F/CPTFE is close to that expected for PTFE (about 2). By the annealing at 300°C, the ratio decreases to 1.7 for both layer thicknesses. The decrease may be due to reorientation of polar C-F groups induced by thermal treatment. Oxygen detected in the samples may result from oxygen incorporation during gold sputtering which may be accompanied by partial degradation and oxidation of PTFE macromolecular chain or degradation products. Subsequent annealing leads to reorientation of the oxidized groups toward the sample bulk and corresponding decrease of the surface concentration of oxygen. The same effects have been observed earlier on plasma-modified polyolefines [64]. It is also evident from Table 1 that annealing causes resorption of contamination carbon both hydrogenated and oxidized one [65]. Changes in the morphology of the gold layer after the annealing are manifested in changes of the gold and fluorine concentrations as observed in XPS spectra. After the annealing, the observed gold concentration decreases and fluorine concentration increases dramatically, these changes clearly indicate formation of isolated Au islands similar to those in case of Au-coated glass substrate [16].

**Table 1.** Atomic concentrations (in at. %) of C (1s), O (1s), Au (4f) and F(1s) in Au sputtered PTFE samples with Au effective thickness 20 and 80 nm after deposition (RT) a after annealing (300°C) measured by XPS. CPTFE represents calculated concentration from XPS data of carbon (in at. %) originating from PTFE only, F/CPTFE stands for fluorine to PTFE carbon ratio [63].

| Au layer Thickness | Temperature | Atomic concentrations of elements in at. % | | | | | |
|---|---|---|---|---|---|---|---|
| | | C | $C^{PTFE}$ | O | Au | F | $F/C^{PTFE}$ |
| 20 nm | RT | 43.5 | 4.4 | 6.5 | 41.6 | 8.5 | 1.93 |
| | 300°C | 37.8 | 34.8 | 0.4 | 3.4 | 58.4 | 1.68 |
| 80 nm | RT | 41.0 | 3.1 | 4.4 | 48.6 | 6.0 | 1.94 |
| | 300°C | 36.8 | 27.2 | 1.2 | 14.8 | 47.2 | 1.74 |

## 3.3. Surface Properties and Morphology

Another quantity characterizing the structure of the sputtered gold layers is zeta potential determined from electrokinetic analysis. Dependence of zeta potential on the gold layer thickness for as-sputtered samples (RT) and annealed samples at 300°C is shown in Fig. 16. For as-sputtered samples and very thin gold layers, the zeta potential is close to that of pristine PTFE due to the discontinuous gold coverage since the PTFE surface plays dominant role in zeta potential value. Then, for thicker layers, where the gold coverage prevails over the original substrate surface, the zeta potential decreases rapidly and for the thicknesses above 20 nm remains nearly unchanged, indicating total coverage of original substrate by gold. For annealed samples, the dependence on the layer thickness is quite different. It is seen that the annealing leads to a significant increase of the zeta potential for very thin layers. This increase may be due to thermal degradation of the PTFE accompanied by production of excessive polar groups on the polymer surface, which plays the important role when the gold coverage is discontinuous. Moreover, the surface roughness increases at this moment too (see Table 1 and Fig. 17 below) [66]. Then, for medium thicknesses, ranging from 20 to 70 nm, the zeta potential remains unchanged and finally it decreases again for higher thicknesses due to the formation of continuous gold coverage. It appears that the results of electrokinetic analysis (Fig. 16) and measurement of the sheet resistance (Fig. 13) are highly correlated.

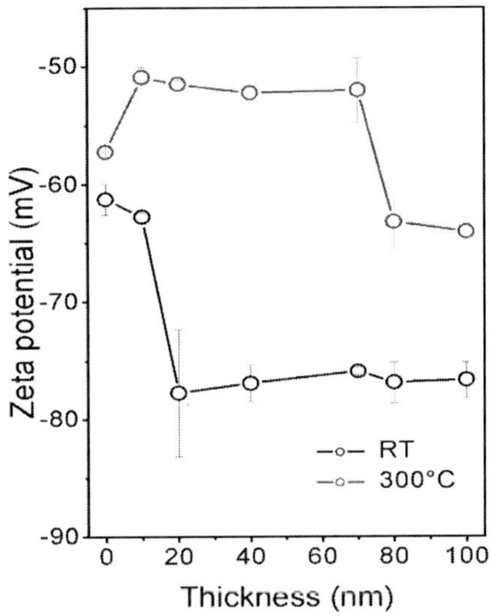


**Figure 16.** Dependence of zeta potential on the Au layer thickness for as-sputtered samples (RT) and the samples annealed at 300°C [63].

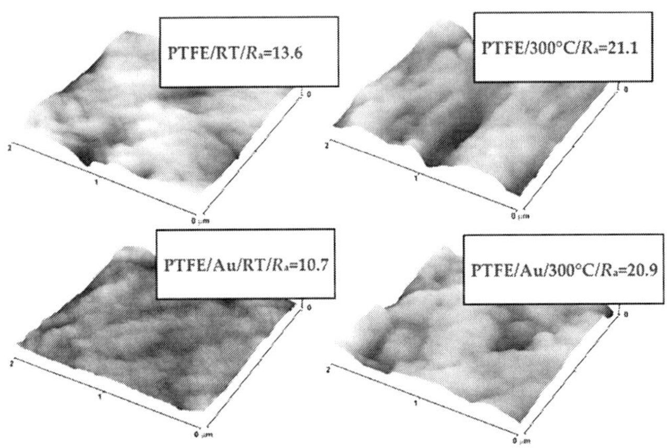


**Figure 17.** AFM images of pristine (PTFE) and Au coated (PTFE/Au) samples (thickness of 20 nm) before (RT) and after annealing at 300°C. Numbers in frames are measured surface roughnesses Ra in nm [63].

The rapid decrease in the sheet resistance occurs at the same layer thickness as the decrease in zeta potential. Both correlated changes are connected with creation of continuous, conductive gold coverage. Another interesting fact is that even for the layers with thicknesses above 80 nm, the values of the zeta potential measured on as-sputtered and annealed samples differ significantly. This can be due to higher fluorine concentration in the annealed samples and the fact that the C-F bond is more polar and exhibits higher wettability. It should be also noted that the value of the zeta potential may be affected by the surface roughness too. In general, it follows that the thicker the gold coverage the lower the zeta potential is, reflecting the electrokinetic potencial of metal itself.

The changes in the surface morphology after the annealing were studied by AFM. AFM scans of pristine and Au-coated (20 nm) samples before and after annealing are presented in Fig. 17. One can see that the annealing causes a dramatic increase in the surface roughness of the pristine polymer. Since the annealing temperature markedly exceeds PTFE glassy transformation temperature (TgPTFE = 126°C) the increase in the surface roughness is probably due to thermally induced changes of PTFE amorphous phase. The gold sputtering leads to a measurable reduction of the sample surface roughness. The reduction may be due to preferential gold growth in holes at the PTFE surface. Annealing of the gold-coated sample leads to significant increase of the surface roughness too. In this case, the increase is a result of both, the changes in the surface morphology of underlying PTFE and the changes in the morphology of the gold layer. After annealing, the surface roughness of pristine and gold-coated samples is practically the same. This finding is in contradiction with similar study accomplished on gold layers deposited on glass substrate [16]. Possible explanation of this fact probably lies in much better flatness of the glass substrate and in lower thermal stability of PTFE substrate during annealing.

# 4. SELF-ORGANIZED GOLD NANOSTRUCTURES

Purpose of this section lies in description of phenomena taking place during both interaction of polarized laser light with the surface of polymeric material and its subsequent coating by metal. It will be shown that modification of the polyethylenetherephtalate (PET) surface with linearly polarized light from pulsed KrF laser has a significant effect on the properties of subsequently deposited gold nanolayers and the choice of the deposition technique is crucial owing to the quality of prepared coatings.

## 4.1. Surface Morphology and Structure Parameters

It has been shown [67] that by the KrF laser irradiation with several thousand of pulses a periodic ripple structure is formed at a PET surface for a fluence range from about 4.2 to 18.8 mJ cm−2. The ripples have a fluence-independent width Λ, which is given by the formula Λ = λ/(1 – sin(θ)), Eq. (1) [68], where λ is wavelength of a laser light used, n the effective refractive index of material, and θ the angle of incidence. Fig. 18 displays AFM images of pristine PET and PET irradiated at different laser fluences. The sample irradiated with the laser fluence of 3.4 mJ cm−2 exhibits a rougher surface than the flat un-irradiated pristine PET. There is a noticeable modulation, although no ripple formation is visible. At higher laser fluences periodic ripple structures have developed in the irradiated area. At a laser fluence of 6.6 mJ cm−2, a regular and uniform coverage of the PET surface with ripples is reached. These results are based on AFM measurements, as those shown in Fig. 18. There is a good correlation between height and surface roughness of ripples over the whole laser fluence range shown in the figure. Both parameters reach the maximum value at a fluence of 6.6 mJ cm−2, which corresponds to a ripple height of about 90 nm.

The height and the roughness of the ripples as a function of the applied laser fluence are shown in Fig. 19.

Fig. 20 shows AFM images of the PET irradiated under different incidence angles of the laser beam. For larger angles of incidence, the spacing between two neighboring ripples is wider. For the incidence angle of 0° and 22.5°, the observed spacing of the ripples is in good agreement with the value calculated by Eq. (1) with an effective index of refraction n ≈ 1.2. The agreement for the incidence angle of 45° is less pronounced. The discrepancy may be due to changes of the polymer refractive index induced by the UV laser irradiation as reported earlier [69].

FIB cuts of laser irradiated and gold coated PET samples were investigated by SEM (see Fig. 21). After sputtering, the gold is deposited in the form of "nanowires", which grow mainly at the ridges of the ripples. The FIB cuts reveal that there could be gaps between the individual wires and that the metal layer may be discontinuous. The width of the gold nanowires directly correlates to the width of the ripples formed before gold deposition. Additionally, a granularity is visible along the wires, but the FIB cut images suggest that the grains may be interconnected. The morphology of the gold layers deposited by evaporation is distinctly different. The gold is also deposited in the valleys of the ripple structure.

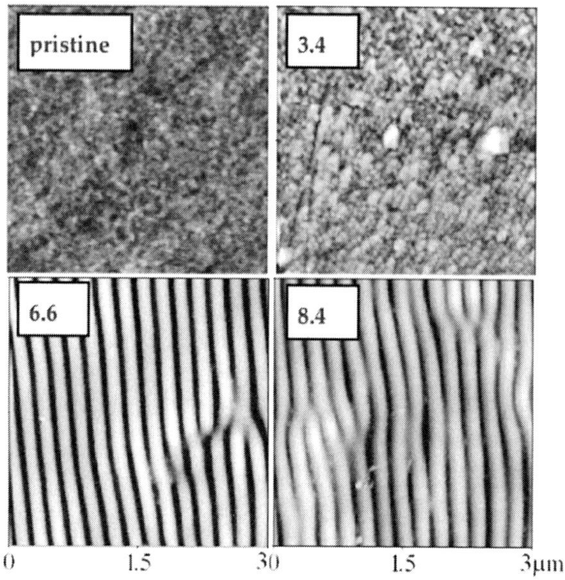


**Figure 18.** AFM images of the PET irradiated at different KrF laser fluences; the numbers in the inset refer to the laser fluence in mJ cm−2 employed for irradiation of the PET foils, while pristine stands for unirradiated pristine PET [70].

The nanowire structure of the sputtered gold layer can be observed also after gold deposition onto PET samples irradiated by the laser under an angle of incidence of 45°. Again, the evaporation of gold leads to a continuous coverage copying nanostructured polymer surface. The reason for the different observed gold morphologies after sputtering (nanowires) and evaporation (homogeneous gold coverage) is still unclear. The different particle energies in both processes are one possible reason. For sputtering, the particle energy may be considerably higher because of sample charging effect, while the evaporated materials should be slower (i.e., colder) and closer to thermodynamical equilibrium. The electrical charge of the sputtered particles can have also direct influence on layer formation, while the evaporated material should be mainly neutral. Other reasons may be the different deposition rates, which were a factor of two lower for sputtering than for evaporation, and possible differences of the substrate and gold layer temperature during the deposition in the two different techniques.

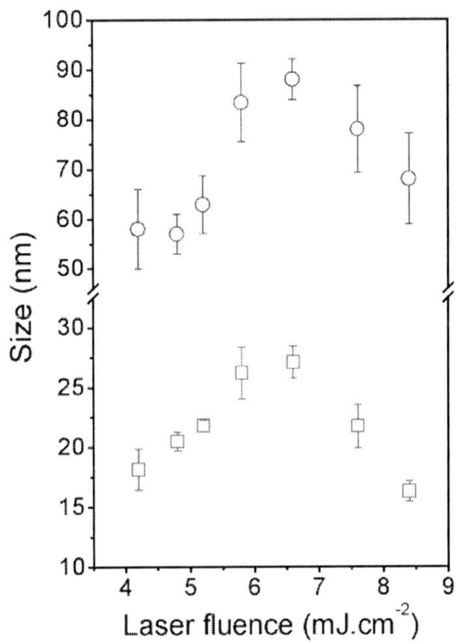


**Figure 19.** Dependence of the ripple height (○) and roughness (□) on the KrF laser fluence employed for the PET irradiation [70].

## 5. SUMMARY

In summary, this chapter gives a comprehensive insight into the problematic of ultrathin gold films formed by physical deposition techniques on glass and polymeric substrates. Particular emphasis is given to the processes taking place during post-deposition annealing of prepared layers. In the case of glass substrate, the sputtering times and the layer effective thicknesses were chosen to span the region of the transition from discontinuous to continuous gold layer. For short sputtering times electrically discontinuous layers are obtained comprising discrete gold crystallites. The crystallite size in the as-sputtered samples is a monotonously increasing function of the sputtering time. The dependence of the lattice parameter of the gold crystallites forming the layer on the sputtering time is rather complicated with a rapid increase for shorter sputtering times and subsequent decrease for longer sputtering times. The decrease can be explained by relaxation processes in the thicker layers. The annealing has significant influence on the properties of the gold layers. For the annealed samples the lattice parameter practically does not depend on the sputtering time. The crystallite size first increases rapidly up to the sputtering time of 250 s, achieves a maximum and then decreases. The electrical properties (concentration, mobility of charge carriers and volume resistivity) and NIR optical properties of the gold layers change dramatically as the function of the sputtering time. Other significant changes, especially for electrically discontinuous layers, are observed as a result of the annealing. This is probably due to the different mechanism of free charge carrier transport, where also the quantum surface effects could be present in case of observed island structure mainly after annealing. For electrically continuous layer the concentration and the mobility are invariable. Similar behaviour exhibits also gold layers on polymeric substrate. From the measurement of the sheet resistance the transition from discontinuous to continuous gold coverage was found at the layer thicknesses of 10-15 nm for as-sputtered samples. After annealing at 300°C the transition point increases to about 70 nm, the increase indicating substantial rearrangement of the gold layer. The rearrangement is confirmed also by XPS measurement and an electrokinetic analysis. By XPS measurement contamination of the gold coated PTFE samples with carbon and the presence of oxidized structures, created during gold sputtering were proved. The annealing results in significant increase of the surface roughness of both pristine and gold sputtered PTFE.

Modification of the PET surface with linearly polarized light from pulsed KrF laser has a significant effect on the properties of subsequently deposited gold nanolayers and the choice of the deposition technique is crucial owing to the quality of prepared coatings. Subsequent deposition of 200 nm thick gold layer caused a decrease of the surface roughness. While by evaporation a continuous metal coverage is formed, copying nanostructured polymer surface, in the case of sputtering a nanowire-like structure of the gold coating can be observed. It was shown that the width of the nanowires can be tailored by the width of the ripples formed by preceding laser irradiation. We demonstrate a technique for the controlled patterning of polymer surfaces, including the creation of nanopatterned, regular gold structures (nanowires). In principle, this technique could be employed for the creation of metal-polymer composites with interesting electrical, mechanical, and optical properties, which could find novel applications in micro and nanotechnology.

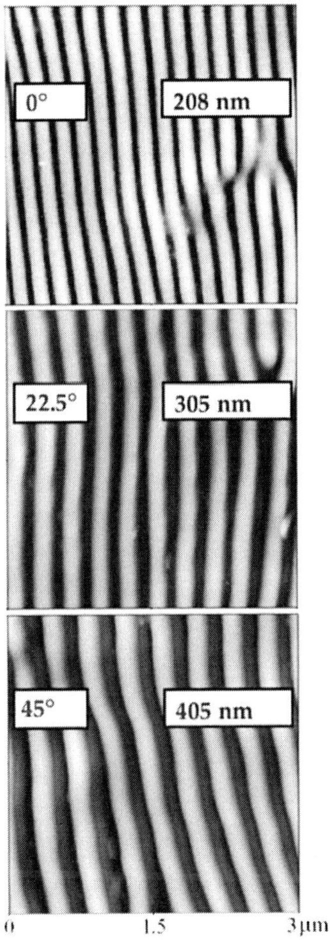


**Figure 20.** AFM images of the PET irradiated at a KrF laser fluence of 6.6 mJ cm−2 under the different incidence angle of laser beam (0, 22.5 and 45°). The numbers in the insets in the upper left corner refer to the angle of incidence of the laser beam and in the insets in the upper right corner to the ripple period in nm [70].

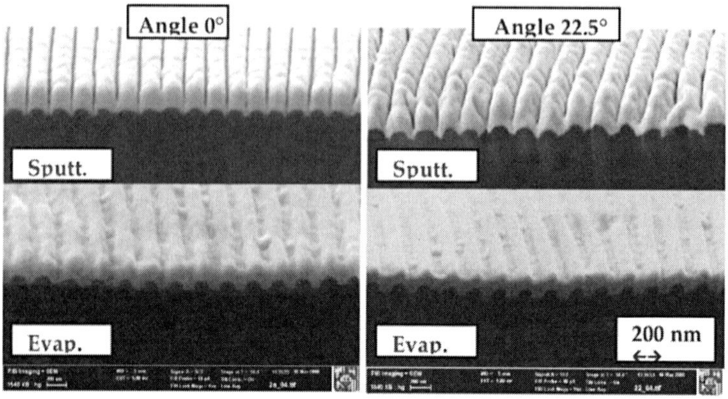


**Figure 21.** FIB-SEM images of the gold coatings on PET samples irradiated by a KrF laser under incidence angles 0 and 22.5° (fluence 6.6 mJ cm−2). The gold deposition was performed either by sputtering (Sputt.) and evaporation (Evap.) [70].

## ACKNOWLEDGEMENT

Financial support of this work from the GACR projects No. P108/11/P337, P108/10/1106 and 106/09/0125 is gratefully acknowledged.

## REFERENCES

1. Rao CNR, Kulkarni GU, Thomas PJ, Edwards PP2002Size-Dependent Chemistry: Properties of NanocrystalsChem. eur. j. 82539

2. E. Roduner, 2006Size Matters: Why Nanomaterials are DifferentChem. soc. rev. 35583592

3. M. C. Daniel, D. Astruc, 2004Gold Nanoparticles: Assembly, Supramolecular Chemistry, Quantum-Size-Related Properties, and Applications Toward Biology, Catalysis, and Nanotechnology.Chem. rev. 104293346

4. S. Guoa, E. Wang, 2007Synthesis and Electrochemical Applications of Gold Nanoparticles. Anal. chim. acta 598181192

5. T. Yonezawa, T. Kunitake, 1999Preparation of Anionic Mercapto Ligand-Stabilized Gold Nanoparticles and their Immobilization. Colloid surf. A 149193199

6. J. Turkevich, P. C. Stevenson, J. Hillier, 1951A Study of the Nucleation and Growth Processes in the Synthesis of Colloidal GoldDiscuss faraday soc. 115559

7. Kim HJ, Jung SM, Kim BJ, Yoon TS, Kim YS, Lee HH2010Characterization of Charging Effect of Citrate-Capped AuNP Pentacene Device. J. ind. eng. chem. 16848851

8. S. P. Xu, B. Zhao, W. Q. Xu, Y. G. Fan, 2005Preparation of Au-Ag Coreshell Nanoparticles and Application of Bimetallic Sandwich in SERS. Colloid surf. A 257-258: 313-317.

9. J. C. Sánchez-López, Kolodziejczyk. L. Abad, E. Guerrero, A. Fernández, 2011Surface-Modified Pd and Au NPs for Anti-Wear Applications. Tribol. int. 44720726

10. S. Seino, T. Kinoshita, Y. Otome, T. Maki, T. Nakagawa, K. Okitsu, Y. Mizukoshi, T. Nakayama, T. Sekino, K. Niihara, T. A. Yamamoto, 2004Gamma-Ray Synthesis of Composite Nanoparticles of Noble Metals and Iron Oxides. Scripta mater. 51467472

11. Kan CX, Zhu XG, Wang GH2006Single-Crystalline Gold Microplates: Synthesis, Characterization, and Thermal Stability.J. phys. chem. B 11046514656

12. H. B. Liu, J. A. Ascencio, M. Perez-Alvarez, Yacaman, 2001Melting Behavior of Nanometer Sized Gold IsomersJ. surf. sci. 4918898

13. J. Siegel, O. Lyutakov, V. Rybka, Z. Kolská, V. Švorčík, 2011Properties of Gold Nanostructures Sputtered on Glass. Nanoscale res. lett. 6: 96.

14. V. Švorčík, J. Siegel, P. Šutta, J. Mistrík, P. Janíček, P. Worsch, Z. Kolská, 2011Annealing of Gold Nanostructures Sputtered on Glass Substrate. Appl. phys. A 102605610

15. C. Solliard, M. Flueli, 1985Surface Stress and Size Effect on the Lattice-Parameter in Small Particles of Gold and Platinium. Surf. sci. 156487494

16. V. Švorčík, O. Kvítek, O. Lyutakov, J. Siegel, Z. Kolská, 2011Annealing of Sputtered Gold Nano-Structures. Appl. phys. A 102747751

17. Heo DN, Yang DH, Moon HJ, Lee JB, Bae MS, Lee SC, Lee WJ, Sun IC, Kwon IK2012Gold Nanoparticles Surface-Functionalized with Paclitaxel Drug and Biotin Receptor as Theranostic Agents for Cancer TherapyBiomaterials33856866

18. S. Lee, H. Chon, S. Y. Yoon, E. K. Lee, S. I. Chang, D. W. Lim, J. Choo, 2012Fabrication of SERS-Fluorescence Dual Modal Nanoprobes and Application to Multiplex Cancer Cell ImagingNanoscale4124129

19. K. Y. Choi, G. Liu, S. Lee, X. Y. Chen, 2012Theranostic Nanoplatforms for Simultaneous Cancer Imaging and Therapy: Current Approaches and Future PerspectivesNanoscale4330342

20. A. C. Curry, M. Crow, A. Wax, 2008Molecular Imaging of Epidermal Growth Factor Receptor in Live Cells with Refractive Index Sensitivity Using Dark-Field Microspectroscopy and Immunotargeted Nanoparticles.J. biomed. opt. 13: 014022 EOF

21. V. Švorčík, Z. Kolská, O. Kvítek, J. Siegel, A. Řezníčková, P. Řezanka, K. Záruba, 2011Soft and Rigid" Dithiols and Au Nanoparticles Grafting on Plasma-Treated Polyethyleneterephthalate. Nanoscale res. lett. 6: 607.

22. V. Švorčík, A. Chaloupka, K. Záruba, V. Král, O. Bláhová, A. Macková, V. Hnatowicz, 2009Deposition of Gold Nano-Particles and Nano-Layers on Polyethylene Modified by Plasma Discharge and Chemical Treatment. Nucl. instrum. meth. B 26724842488

23. Wang ZL2000Characterizing the Structure and Properties of Individual Wire-Like NanoentitiesAdv. mater. 1212951298

24. Hu JT, Odom TW, Lieber ChM1999Chemistry and Physics in One Dimension: Synthesis and Properties of Nanowires and Nanotubes Acc. chem. res. 32435445

25. S. Hollensteiner, E. Spiecker, C. Dieker, W. Jäger, R. Adelung, L. Kipp, M. Skibowski, 2003Self-Assembled Nanowire Formation During Cu

Deposition on Atomically Flat Vse(2) Surfaces Studied by Microscopic Methods. Mater. sci. eng. C 23171179

26. S. Milenkovic, T. Nakayama, M. Rohwerder, A. W. Hassel, 2008Structural characterisation of gold nanowire arraysJ. cryst. growth 311194199

27. Lide DR1994The Handbook of Chemistry and Physics 74th edn. Boca Raton: Chemical Rubber Company. 68 p.

28. Greenwood NN1984Earnshaw A Chemistry of the Element. New York: Pergamon. 126 p.

29. Kouklin NA, Kim WE, Lazareck AD, Xu JM2005Carbon Nanotube Probes for Single-Cell Experimentation and Assays. Appl. phys. lett. 87: 173901 EOF

30. I. Obataya, C. Nakamura, S. Han, N. Nakamura, J. Miyake, 2005Nanoscale Operation of a Living Cell Using an AFM with a Nanoneedle. Nano lett. 52730

31. Gross GW, Wen WY, Lin JW1985Transparent I Electrode Patterns for Extracellular, Multisite Recording in Neuronal Cultures. J. neurosci. meth. 15243252

32. J. Pine, 1980Recording Action-Potentials from Cultured Neurons with Extracellular Micro-Circuit Electrodes. J. neurosci. meth. 21931

33. V. Švorčík, N. Kasálková, P. Slepička, K. Záruba, L. Bačáková, M. Pařízek, V. Lisa, T. Ruml, A. Macková, 2009Cytocompatibility of Ar(+) Plasma Treated and Au Nanoparticle-Grafted PE. Nucl. instrum. meth. B 26719041910

34. Chithrani BD, Ghazani AA, Chan WCW2006Determining the Size and Shape Dependence of AuNP Uptake into Mammalian Cells. Nano lett. 6662668

35. Rosi NL, Giljohann DA, Thaxton CS, Lytton-Jean AKR, Han MS, Mirkin CA2006Oligonucleotide-Modified Gold Nanoparticles for Intracellular Gene Regulation. Science 31210271030

36. F. Favier, E. C. Walter, M. P. Zach, T. Benter, R. M. Penner, 2001Hydrogen Sensors and Switches from Electrodeposited Pd Mesowire Arrays. Science 29322272231

37. Q. Wan, Q. H. Li, Y. J. Chen, T. H. Wang, X. L. He, J. P. Li, C. L. Lin, 2004Fabrication and Ethanol Sensing Characteristics of ZnO Nanowire Sensors. Appl. phys. lett. 8436543656

38. F. Yang, D. K. Taggart, R. M. Penner, 2009Fast, Sensitive Hydrogen Gas Detection Using Single Palladium Nanowires That Resist Fracture. Nano lett. 921772182

39. B. He, T. J. Morrow, C. D. Keating, 2008Nanowire Sensors for Multiplexed Detection of BiomoleculesCurr. opin. chem. biol. 12522528

40. Y. Cui, Q. Q. Wei, H. K. Park, C. M. Lieber, 2001Nanowire Nanosensors for Highly Sensitive and Selective Detection of Biological and Chemical SpeciesScience29312891292

41. V. V. Poborchii, T. Tada, T. Kanayama, 2002Photonic-Band-Gap Properties of Two-Dimensional Lattices of Si NanopillarsJ. appl. phys. 9132993305

42. F. C. K. Au, K. W. Wong, Y. H. Tang, Y. F. Zhang, I. Bello, S. T. Lee, 1999Electron Field Emission from Silicon Nanowires. Appl. phys. lett. 7517001702

43. W. Gotschy, K. Vonmetz, A. Lietner, F. R. Aussenegg, 1996Thin Films by Regular Patterns of Metal Nanoparticles: Tailoring the Optical Properties by NanodesignAppl. phys. B 63381384

44. T. W. H. Oates, A. Keller, S. Facsko, A. Mücklich, 2007Aligned Silver Nanoparticles on Rippled Si Templates with Anisotropic Plasmon Absorption. Plasmonics 24750

45. Busbee BD, Obare SO, Murphy CJ2003An Improved Synthesis of High-Aspect-Ratio Gold NanorodsAdv. mater. 15414416

46. X. Y. Zhang, L. D. Zhang, Y. Lei, L. X. Zhao, Y. Q. Mao, 2001Fabrication and Characterization of Highly Ordered Au Nanowire ArraysJ. mater. chem. 1117321734

47. J. K. N. Mbindyo, T. E. Mallouk, J. B. Mattzela, I. Kratochvilova, B. Razavi, T. N. Jackson, T. S. Mayer, 2002Template Synthesis of Metal Nanowires Containing Monolayer Molecular Junctions.J. am. chem. soc. 12440204026

48. M. Wirtz, C. R. Martin, 2003Template-Fabricated Gold Nanowires and NanotubesAdv. mater. 15455458

49. Bartlett. P. N. Ghanem, P. De Groot, A. Zhokov, 2004A Double Templated Electrodeposition Method for the Fabrication of Arrays of Metal NanodotsElectrochem. commun. 6447453

50. V. Švorčík, P. Slepička, J. Švorčíková, M. Špírková, J. Zehentner, V. Hnatowicz, 2006Characterization of Evaporated and Sputtered Thin Au Layers on Poly(Ethylene Terephtalate). J. appl. polym. sci. 9916981704

51. G. Kaune, Metwalli. E. Ruderer, W. Wang, S. Couet, K. Schlage, R. Röhlsberger, S. V. Roth, P. Müller-Buschbaum, 2009In Situ GISAXS Study of Gold Film Growth on Conducting Polymer Films. Appl. mater. interf. 1353362

52. Z. Kolská, V. Švorčík, J. Siegel, 2010Size-Dependent Density of Gold Nano-Clusters and Nano-Layers Deposited on Solid Surface. Collect. czech. chem. C. 75517525

53. I. Donor-Mor, Z. Barkay, N. Filip-Granit, A. Vaskevich, I. Rubinstein, 2004Ultrathin gold island films on silanized glass. Morphology and optical properties1634763483

54. K. Haupl, M. Lang, P. Wissmann, 1986X-Ray-Diffraction Investigations on Ultra-Thin Gold-Films. Surf. interface anal. 92730

55. I. Shyjumon, M. Gopinadhan, O. Ivanova, M. Quaas, H. Wulff, Hippler. R. Helm, 2006Structural Deformation, Melting Point and Lattice Parameter Studies of Size Selected Silver ClustersEur. phys. j. D37409415

56. V. L. Santos, D. Lee, J. Seo, F. L. Leon, D. A. Bustamante, S. Suzuki, Y. Majima, T. Mitrelias, A. Ionescu, C. H. W. Barnes, 2009Crystallization and Morphology of Au/SiO(2) Thin Films Following Furnace and Flame Annealing. Surf. sci. 60329782975

57. J. Tauc, 1974Amorphous and Liquid Semiconductors.Heidelberg: Springer. 202 p.

58. W. Fischer, H. Geiger, P. Rudolf, P. Wissmann, 1977Structure Investigations on Single-Crystal Gold-FilmsAppl. phys. 13245253

59. P. Slepička, Z. Kolská, J. Náhlík, V. Hnatowicz, V. Švorčík, 2009Properties of Au Nanolayers on PET and PTFE. Surf. interface anal. 41741745

60. M. Hovel, B. Gompf, M. Dressel, 2010Dielectric Properties of Ultrathin Metal Films Around the Percolation ThresholdPhys. rev. B 81: 035402.

61. Hodgman CD1975Handbook of Chemistry. Cleveland:Chemical Rubber. 264 p.

62. K. Chopra, 1969Thin Film PhenomenaNew York: Wiley. 138 p.

63. J. Siegel, R. Krajcar, Z. Kolská, V. Hnatowicz, V. Švorčík, 2011Annealing of Gold Nanostrucutres Sputtered on Polytetrafluoroethylene. Nanoscale res. lett. 6:588.

64. V. Švorčík, V. Kotál, J. Siegel, P. Sajdl, A. Macková, V. Hnatowicz, 2007Ablation and Water Etching of PE Modified by Ar Plasma. Polym. degrad. stabil. 9216451649

65. J. Siegel, A. Řezníčková, A. Chaloupka, P. Slepička, V. Švorčík, 2008Ablation and water etching of plasma-treated polymers. Radiat. eff. deffect S 163779788

66. V. Švorčík, A. Řezníčková, Z. Kolská, P. Slepička, 2010Variable Surface Properties of PTFE Foils. e-Polymers 13316

67. J. Siegel, P. Slepička, J. Heitz, Z. Kolská, P. Sajdl, V. Švorčík, 2010Gold Nano-Wires and Nano-Layers at Laser-Induced NanoRipples on PET. Appl. surf. sci. 25622052209

68. D. Bäuerle, 2000Laser Processing and ChemistryBerlin-Heidelberg-New York: Springer-Verlag. 378 p.

69. Dunn DS, Ouderkirk AJ1990Chemical and Physical Properties of Laser-modified PolymersMacrolmolecules 23770774

70. J. Siegel, J. Heitz, V. Švorčík, 2011Self-organized Gold Nanostructures on Laser Patterned PET. Surf. coat. technol. 206517521

5

# CHARACTERIZATION OF PRECIPITATES IN A MICROALLOYED STEEL USING QUANTITATIVE X-RAY DIFFRACTION

**J. Barry Wiskel [1,*], Junfang Lu [2], Oladipo Omotoso [3], Douglas G. Ivey [1] and Hani Henein [1]**

[1]Department of Chemical and Materials Engineering, University of Alberta, Edmonton, AB T6G 2V4, Canada

[2]Enbridge Pipelines, 10201, Jasper Ave, Edmonton, AB T5J 3N7, Canada

[3]Suncor Energy, Suncor Energy Centre, 111, 5 Ave SW, Calgary, AB T2P 3E3, Canada

## ABSTRACT

Quantitative X-ray diffraction (QXRD) (also known as the Rietveld method) was used to analyze the precipitates present in Grade 100 microalloyed steel. The precipitates were extracted from the steel using electrolytic dissolution and the residue from the dissolution was analyzed using XRD. The XRD pattern exhibited three (3) distinct diffraction peaks, and significant broadening of a fourth peak corresponding to the <10 nm size precipitates. QXRD analysis was applied to the XRD pattern to obtain precipitate size, composition, and weight fraction data for each of the four diffraction peaks observed. The predicted mean precipitate diameter and average atomic composition of the nano-size (<10 nm) precipitates was 4.7 nm and $(Nb_{0.50}Ti_{0.32}Mo_{0.18})(C_{0.59}N_{0.41})$, respectively. The predicted precipitate size correlates well with the average size of precipitates measured in previous work by the authors using both transmission electron microscopy (TEM) and small angle neutron scattering (SANS). The average atomic composition correlates well with the composition measured in this work using energy dispersive X-ray (EDX) analysis of individual nano-sized precipitates. The calculated weight fraction of the nano-size precipitates in the extracted residue was 42.2 wt. %. The calculated atomic compositions of the other three diffraction peaks were TiN, $(Ti_{0.87}Nb_{0.13})N$, and $(Nb_{0.82}Ti_{0.18})(C_{0.87}N_{0.13})$ with weight fraction values of 12.9 wt. %, 31.7 wt. %, and 13.1 wt. %, respectively. The sizes of both the $(Ti_{0.87}Nb_{0.13})N$ and the $(Nb_{0.82}Ti_{0.18})(C_{0.87}N_{0.13})$ groups of precipitates were directly measured and were observed to range from 150 nm to 570 nm and from 90 nm to 475 nm, respectively. QXRD was unable to determine a reasonable mean precipitate size for either of these two groups of precipitates. The wide compositional range (*i.e.*, varying levels of Nb and Ti) of these precipitates (as measured by EDX) resulted in XRD peak broadening that was erroneously interpreted as a size broadening effect.

## KEYWORDS

microalloyed steel; quantitative X-ray diffraction; precipitates; size; composition

## 1. INTRODUCTION

The characterization of precipitates in microalloyed steels can be undertaken using a number of analytical techniques, including optical microscopy [1,2], scanning electron microscopy (SEM) [3,4,5,6], transmission electron microscopy (TEM) [2,3,7,8], energy dispersive X-ray (EDX) analysis [2,3,8,9], and small angle neutron scattering (SANS) [10]. Each technique has its own advantages (and disadvantages) in terms of quantifying the number density, volume or weight fraction, composition and/or size of the precipitates. SANS is advantageous in that a larger number of precipitates are examined simultaneously compared with SEM and TEM, where a small number of individual precipitates are sequentially analyzed. Due to the wide range of precipitate sizes (from several microns down to the nanometer scale

[2,11,12]) present in microalloyed steels that have undergone thermo-mechanical controlled processing (TMCP), and the relatively low volume fractions and diverse compositions of the precipitates (e.g., TiN, $(Ti_x,Nb_{1-x})(C_yN_{1-y})$, *etc.*), most, if not all, of the above mentioned characterization techniques are concurrently used.

This paper outlines the application of quantitative X-ray diffraction (QXRD) to characterize precipitates chemically extracted from Grade 100 microalloyed steel. The precipitate characteristics quantified by QXRD include precipitate size, weight fraction and average atomic composition. The extracted precipitate sample size of 100 mg represents a significant number of precipitates being simultaneously examined. The precipitate characteristics obtained from QXRD are compared with the composition and size of a select number of precipitates measured with EDX and TEM and from previous analysis on nano precipitate sizing conducted (on the same Grade 100 steel) by the authors using TEM and SANS.

## 2. BACKGOUND

The Rietveld method, in conjunction with X-ray diffraction (XRD)—commonly referred to as quantitative XRD (QXRD) is a characterization methodology that has been applied to many metallurgical and materials systems to measure phase fractions [13], grain size in polycrystalline materials [14] and the average size of powder materials [15,16]. It is this latter application that has instigated the application of QXRD to the characterization of precipitates in microalloyed steel. This background section will review the general QXRD methodology with special emphasis on the quantification of precipitate size.

### 2.1. Qxrd

QXRD [17] mathematically calculates an XRD pattern (*i.e.*, intensity *versus* scattering angle 2θ) for a material and compares the calculated pattern with a measured diffraction pattern in a process known as the fundamental parameters approach (FPA). The calculated diffraction profile is obtained from parameters describing both the diffractometer (*i.e.*, the instrument) and the material being analyzed [18]. The latter group of parameters includes the crystallographic structure of the phase(s) presents (*i.e.*, crystal structure, atom type, and ordering) and specific phase characteristics such as the diffraction domain size (analogous to grain size or individual precipitate size for very fine precipitates) and internal microstrain. A refinement of these variables is performed until a "best fit" is achieved between the entire observed diffraction pattern and the predicted pattern [16] as per the following equation:

$$R = \sum_i w_i (I_{iobs} - I_{ical})^2 \tag{1}$$

where $R$ is the value to be minimized, $I_{iobs}$ and $I_{ical}$ are the observed and calculated XRD intensities, respectively (at any angle θ) and $w_i$ is the statistical weight of each observation point, which takes into account the accuracy of the diffraction experiment.

The intensity of the calculated diffraction profile ($I_{cal}$) for a single phase material can be determined using the following equation [18]:

$$I_{cal} = S \sum_K L_K |F_K|^2 \phi (2\theta_i - 2\theta_K) P_K A + I_b \tag{2}$$

where $S$ is a scale factor (used in the calculation of weight fraction of each phase in a multi-phase system), $L_K$ contains the Lorentz polarization and multiplicity factors, $F_K$ is the structure factor which accounts for the fractional atom locations, the Miller indices (*hkl*) of the plane and the atomic scattering factor of the atom type, $P_K$ is a preferred orientation function, $A$ is an absorption factor specific to the atom type and the X-ray source, $I_b$ is the background intensity and φ is a profile function used to incorporate the effect of both domain size and microstrain on diffraction intensity at different angles ($\theta_i$) relative to the Bragg diffraction angle ($\theta_K$). of importance to the work presented in this paper is the profile function φ [19,20] used to determine domain size (and microstrain), the scale factor for weight fraction and variations in both the absorption factor and atomic scattering factor for each atom (e.g., Nb, Ti, *etc.*) that allows for the determination of precipitate atomic composition.

## 2.2. Domain Size

In a physical sense, diffraction domain size is a measurement of the depth (perpendicular to the diffraction plane) of a continuous set of coherent planes that contribute to Bragg diffraction. In materials with relatively large domain sizes, the diffraction peak decays rapidly on either side of the Bragg angle due to destructive wave interference between X-rays scattering from the near surface plane of atoms and those scattering from a relatively distant plane in the material. The net result is a relatively narrow diffraction peak where any broadening in the diffraction peak is attributed to instrumental effects. However, as domain size falls below approximately 100 nm [21], the diffraction peak begins to noticeably broaden due to the incomplete annihilation of the diffracted X-rays on either side of the Bragg peak. This reduced wave annihilation results in a measureable broadening of the peak beyond any instrumental broadening present.

In an "idealized" single crystal precipitate, the domain size would represent a physical measurement of the precipitate size [15]. It should be noted that domain size does not necessarily translate into precipitate size but represents (for an integral breadth analysis) the volume average mean column height ($D_v$) in the direction normal to the reflecting planes [22]. In addition, broadening of the peak due to domain size depends not only on the dimension normal to the diffracting planes but also on the geometric shape [23] of the diffracting planes in three dimensions (e.g., cube *vs.* sphere) [24]. As a further complication, there is not a specific single domain size but typically a distribution of sizes which can complicate the broadening effect and, hence, render the value of $D_v$ as an average value only.

For polycrystalline spherical precipitates, the presence of crystalline defects such as dislocation arrays (small angle boundaries), stacking faults, twins and/or high angle grain boundaries [23] would delineate the precipitate into multiple domains. The presence of these multiple domains precludes a direct measurement of the precipitate size from diffraction broadening.

## 2.3. Microstrain

Microstrain (or inhomogeneous strain) within the crystal lattice arises from local atomic positional distortions due to the presence of defects such as dislocations, solid solution elements, and vacancies [25]. The greater the number of defects/homogeneities in the material, the greater the microstrain within the domain and, hence, the greater the peak broadening. Microstrain should not be confused with macrostrain (or residual stress), which is a global material phenomenon, that translates into either an increase or decrease in the plane spacing depending upon the direction and magnitude of the residual stress. As both domain size and microstrain result in peak broadening their individual effects must be separated. This can be undertaken by observing the variation in peak broadening as a function of the diffracting angle (θ) (e.g., a Williamson-Hall Plot) or via the use of appropriate profile functions (Gaussian and/or Lorentzian functions), which delineate the individual effects.

## 2.4. Profile Functions

Gaussian (G) and Lorentzian (L) profile functions are typically used to describe diffraction peak broadening resulting from both a small domain size and microstrain and have the following respective forms [18]:

$$I_G = I_o \exp\left\{-\ln 2\left(\frac{2\theta_i - 2\theta_K}{\Gamma_{GK}/2}\right)^2\right\} \tag{3}$$

$$I_L = I_o\left\{1 + \left(\frac{2\theta_i - 2\theta_K}{\Gamma_{LK}/2}\right)^2\right\} \tag{4}$$

where $I_G$ is the intensity for a Gaussian profile, $I_L$ is the intensity for a Lorentzian profile, $I_o$ is the initial intensity and $\Gamma_K$represents the full width at half maximum (FWHM) of any diffraction peak. The value of $\Gamma_K$ is one of several independent variables used in the Rietveld minimization process for the calculation of both domain size and strain. The Gaussian profile is equivalent to a normal distribution [18] while the Lorentzian profile is equivalent to a Cauchy-Lorentz distribution [19]. A major difference between these two functions is that Lorentzian profile has flatter tails (extends for a greater deviation away from the Bragg peak) than a Gaussian distribution exhibiting an equal FWHM.

The application of both the Lorentzian and Gaussian diffraction profiles is based on the experimental usefulness [22,26,27] of both these functions in describing peak broadening and not a fundamental first principles relationship with either domain size or microstrain. To improve the fitting of either (or both) the Gaussian or Lorentzian profile to peak broadening, a combination of the two profiles can be used in a Voigt function [22,25,27] (*i.e.*, broadening due to both domain size and microstrain strain contains both a Lorentzian and Gaussian component [16,25]. The profile function broadening can then be used to calculate the domain size and maximum microstrain using the following equations [16]:

$$\beta_{size} = \frac{\lambda}{L_{Vol} \cdot \cos\theta} \tag{5}$$

$$\beta_{strain} = k \cdot \varepsilon_0 \cdot \tan\theta \tag{6}$$

where $\beta_{size}$ is the integral breadth (a function of $\Gamma_K$) of a Voigt function for broadening due to domain size, $\beta_{srain}$ is the integral breadth of a Voigt function for broadening due to microstrain, $L_{vol}$ is the volume average crystal size (domain size) and $\varepsilon_o$ is the maximum lattice strain. The values of $L_{Vol}$ and $\varepsilon_o$ will be presented in this work as an estimation of the size and microstrain for the precipitates present in Grade 100 microalloyed steel. Other important parameters obtained from QXRD include the weight fraction, composition, and the lattice parameter of a particular phase. All these will be used in characterizing the precipitates of the Grade 100 steel analyzed in this study.

## 2.5. QXRD Programs

There are a number of available programs for conducting Rietveld analysis on diffraction data; these include GSAS (General Structure Analysis System) and TOPAS [23]. For the purpose of the work presented in this paper, TOPAS Academic Software (4.1, Bruker AXS Inc. Madison, WI, USA, 2004) is used. Calculations of composition, domain size, and microstrain are based on the Double Voigt method [16,23]. To ensure that the Rietveld analysis results in a global minimization, different parameter inclusion schemes, and starting values of the parameters were undertaken.

# 3. EXPERIMENTAL PROCEDURE AND RESULTS

The nominal composition of the Grade 100 TMCP microalloyed steel studied in this work is 0.080 wt. % C, 0.011 wt. % N, 0.30 wt. % Mo, 0.20 wt. % Cr, 0.094 wt. % Nb, 0.060 wt. % Ti, and 0.047 wt. % V. The steel was continuously cast and then rolled to its final thickness via TMCP. Following TMCP, a representative sample from the Grade 100 steel strip was extracted and an electrolytic dissolution technique [3] was used to separate the precipitates from the steel matrix. The residue remaining after dissolution (containing the precipitates) was then analyzed using XRD. QXRD was then applied to the experimental diffraction pattern. In addition, TEM and EDX analysis [3] were conducted on a limited number of precipitates in the residue to quantify both the composition (stoichiometry) and size for subsequent comparison with the QXRD results.

## 3.1. Diffraction Pattern of Grade 100 Residue

The XRD pattern of the Grade 100 precipitate residue was obtained using a Bruker D8 diffractometer(Bruker AXS Inc., Madison, WI, USA) with a Co X-ray source. A magnified portion of the XRD pattern, extending from $2\theta = 65°$ to $75°$ (which corresponds to diffraction from the (220) plane in an Fm-3m crystal system), is shown in Figure 1. Three (3) distinct diffraction peaks are observed in Figure 1 and are labeled as 1, 2, and 3. In addition, asymmetric broadening of peak 1 was also observed (as indicated by an arrow). This asymmetric broadening is attributed to the presence of nano-size precipitates. The data indicates that there are four (4) measurable groups of precipitates present in the Grade 100 steel, Groups 1 to 3, shown in Figure 1, and Group 4 indicated by the asymmetric broadening.

## 3.2. QXRD Refinement

QXRD refinement involves calculating a diffraction pattern based on both the XRD instrument parameters and the characteristic crystallographic variables of each precipitate group (described in the background). Table 1 details the specific XRD instrument parameters (and their values) that were directly inputted into the QXRD refinement. The crystallographic variables refined during QXRD for each precipitate group includes the unit cell lattice parameter (a), wt. % of each precipitate, the atomic fraction of each element, the domain size ($L_{vol}$), and the microstrain ($\varepsilon_o$). All the precipitates have a Na-Cl-type crystal structure.

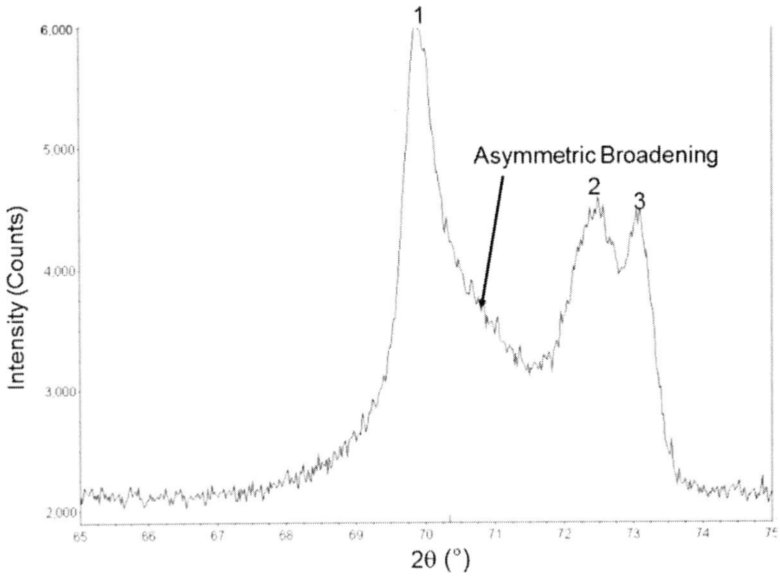


**Figure 1.** Magnified view (65° < 2θ < 75°) of the measured XRD pattern.

**Table 1.** XRD instrument parameters.

| Parameter | Value | Parameter | Value |
|---|---|---|---|
| filament length | 12 mm | primary soller angle | 2.3° |
| sample length | 20 mm | axial beta | 20° |
| receiving slit length | 50 mm | axial del | 0.0053° |
| slit width | 0.15 mm | - | - |

A comparison between the measured (rough line) and calculated diffraction pattern (smooth line) is shown in Figure 2. A good fit (an $R_{wp}$ (weighted profile $R$-factor) = 2.63%) is observed between the measured and predicted profiles and confirms the veracity of the Rietveld refinement in successfully predicting the measured diffraction pattern.

As discussed earlier, the nano-size precipitates do not exhibit a sharp peak similar to the other precipitates (see Figure 1); therefore, their presence must be inferred from shape of the overall diffraction pattern. Figure 3 compares the measured diffraction pattern with the calculated intensity contribution from the Group 4 precipitates for the (220) plane. The asymmetric broadening of the measured diffraction pattern (as indicated by the arrow) corresponds well with the profile calculated for the Group 4 precipitates.

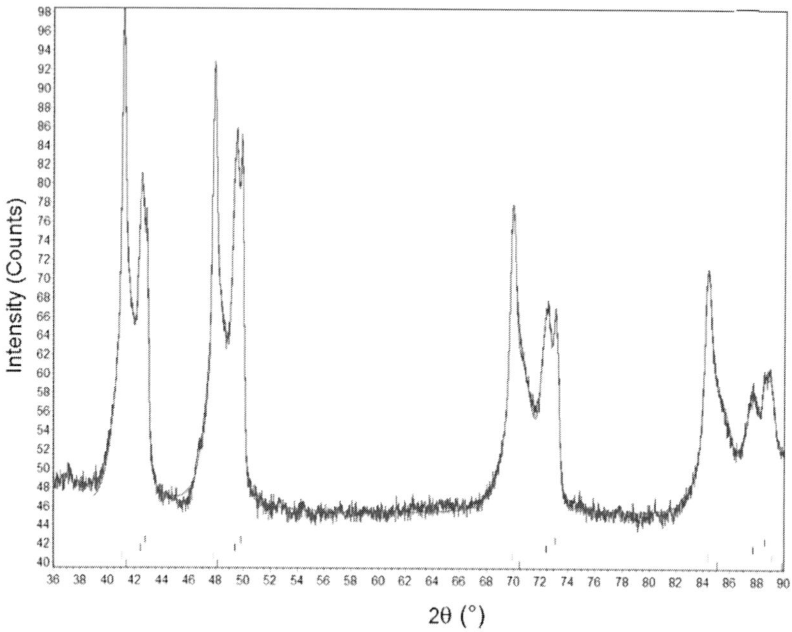


**Figure 2.** Comparison between the measured and predicted XRD pattern.

## 3.3. Predicted Crystallographic and Microstructure Variables

The values of the predicted crystallographic and microstructure variables (*i.e.*, atomic composition, lattice parameter, $L_{vol}$, and $\varepsilon_o$) for all the precipitate groups are summarized in Table 2. The predicted atomic composition of the Group 1 and 2 precipitates show them as primarily titanium nitrides. These nitrides account for 44.6 wt. % of the total number of precipitates collected. Interestingly, both nitride groups also exhibit a relatively low $L_{vol}$ value (70 nm and 115 nm, respectively) and a relatively high internal microstrain (8.8% and 30.1%, respectively). The validity of these predicted values will be discussed later.

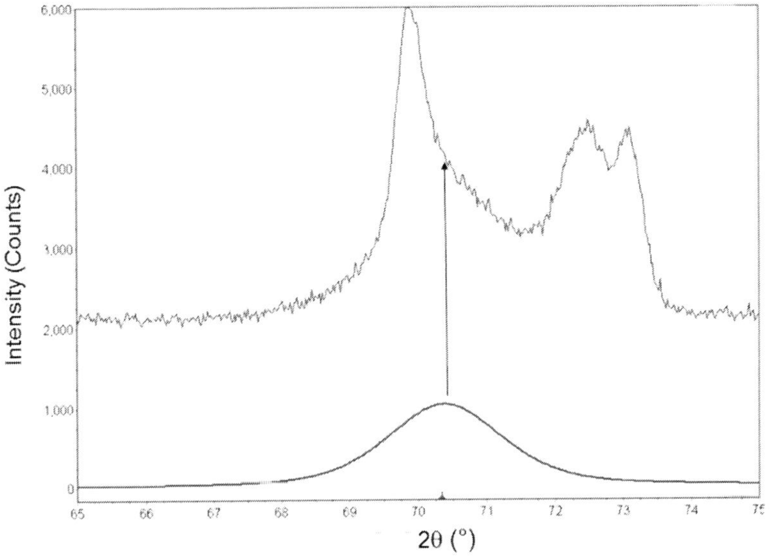


**Figure 3.** Comparison of the measured and predicted (bottom smooth curve) XRD patterns for the Group 4 precipitates for the (220) plane.

**Table 2.** QXRD values for composition, weight fraction, and microstructure dependent variables.

| Group | Atomic Composition | wt. % | $a$ (Å) | $L_{vol}$ (nm) | $\varepsilon_o$ (%) |
|---|---|---|---|---|---|
| 1 | TiN | 12.9 | 4.25 | 70.0 | 8.8 |
| 2 | $Ti_{0.87}Nb_{0.13}N$ | 31.7 | 4.28 | 115.0 | 30.1 |
| 3 | $Nb_{0.82}Ti_{0.18}C_{0.87}N_{0.13}$ | 13.1 | 4.42 | 40.3 | 9.9 |
| 4 | $Nb_{0.50}Ti_{0.32}Mo_{0.18}C_{0.59}N_{0.41}$ | 42.2 | 4.39 | 4.7 | 25.2 |

Alternatively, both Group 3 and 4 precipitates are predicted to be carbonitrides with Nb as the major alloying element in both. The Group 3 precipitates also exhibit a relatively low $L_{vol}$ value (40.3 nm). The Group 4 precipitates were predicted to have a $L_{vol}$ value of 4.7 nm. Given the observed broadening, the small predicted particle size was not unexpected for the latter group. In addition, the Group 4 precipitates were predicted to have Mo present. The small predicted precipitate size and the presence of Mo indicates that the Group 4 precipitates likely arose during the final stages (*i.e.*, relatively low temperature rolling) of TMCP.

## 3.4. EDX/TEM Analysis

The atomic composition of sixty three (63) individual precipitates extracted from the Grade 100 steel residue were obtained through EDX analysis in the TEM [3]. Figure 4 [3] shows an example TEM image (a) and a selected EDX pattern (b) obtained from a nano-size precipitate. The amount of Ti and Nb present in all 63 precipitates analyzed is plotted in Figure 5. For this figure and all subsequent figures, only the metallic elements are considered and the composition does not include the amount of C and N in the precipitates. Included in this figure is a vertical dashed line which subdivides the nano-size precipitates (Group 4) from the Group 1 to 3 precipitates (>90 nm) examined. For the relatively large precipitates, the Ti and Nb compositions vary inversely with each other (*i.e.*, as the amount of Ti decreases the amount of Nb increases). Mo and V were not detected in the Group 1–3 precipitates. The Group 4 precipitates show the amount of Ti to be primarily between 20 at. % and 25 at. %; however the Nb composition varies in a non-regular manner between approximately 38 at. % and 60 at. %. This variation is attributed to the presence of Mo replacing Nb within the carbonitride structure. The presence of Mo and V within the nano-size precipitates will be discussed more fully in subsequent sections.

Although the Nb and Ti compositions for the >90 nm precipitates vary significantly, a number of precipitates have relatively constant Ti compositions. One significant group occurs at ≈82 at. % Ti and to a lesser degree another group at ≈36 at. % Ti. Though a limited number of precipitates were analyzed, the data suggests that these compositions of precipitates are potentially more volumetrically dominant in the steel.

The amount of Ti in the precipitate groups 1–3 (Figure 5) is plotted as a function of measured particle size in Figure 6. Since an exact precipitate size was not determined for the Group 4 precipitates (<10 nm), they have been excluded from this graph. In addition, two very large precipitates, 1770 nm and 3750 nm with compositions of 93 at. % Ti and 89 at. % Ti, respectively, are shown by horizontal arrows as their size markedly exceeds the scale of the graph. In general, precipitates with a composition between 20 and 80 at. % Ti are typically less than 200 nm in size, although several precipitates in the 20–30 at. % Ti range were observed to be significantly larger. Conversely, the precipitates with >80% Ti show a wide range of sizes up to and including 3750 nm in diameter.

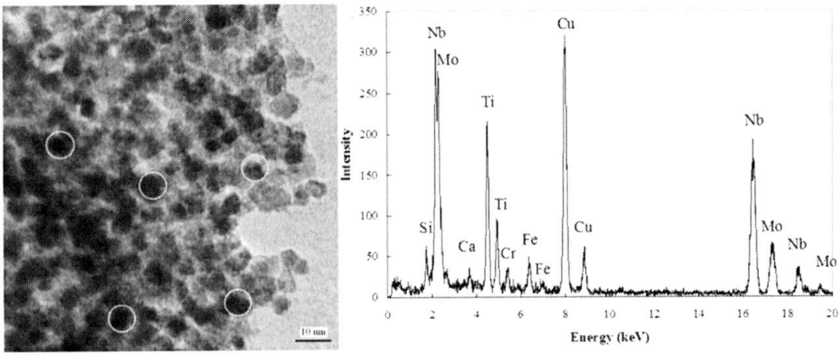


(a) (b)

**Figure 4.** (**a**) TEM BF image of extracted nano-size precipitates and (**b**) EDX spectrum from a single circled region. Note that the Cu peaks are from the support grid, trace amounts of Cr from the Grade 100 composition, while Ca and Si are from inclusions also generated during the extraction process.

The number frequency of precipitates (Groups 2 and 3) as a function of Ti composition is shown in Figure 7. The largest number of the precipitates analyzed contained between 80–90 at. % Ti and 30–40 at. % Ti. The former range corresponds to the Group 2 precipitates, while the latter does not correspond to a specific group detected by QXRD.

# 4. DISCUSSION

The precipitate information (composition and domain size) extracted from QXRD is compared with the composition and size data obtained from the TEM/EDX work undertaken in this paper and with previous work by the authors on precipitate size analysis using TEM and SANS.

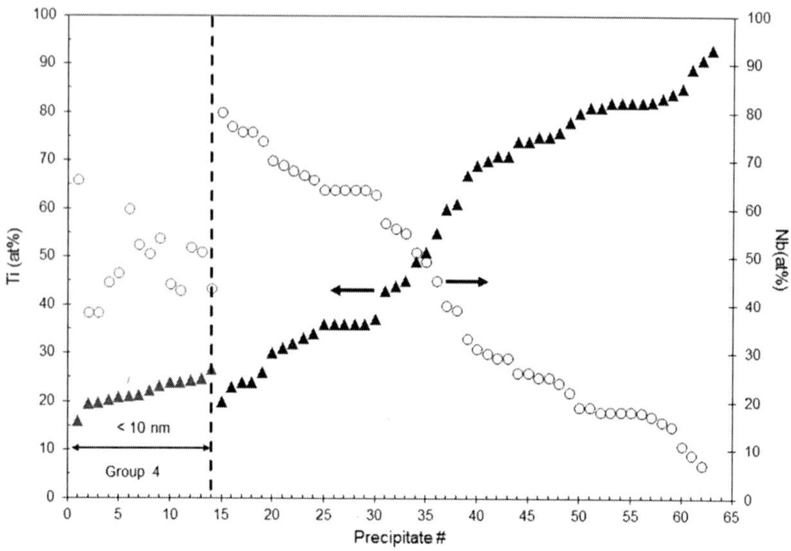


**Figure 5.** Amount of Ti (solid triangles) and Nb (open circles) for selected precipitates measured by EDX analysis.

## 4.1. Composition Analysis

A comparison between the predicted and measured composition of the large (>90 nm) precipitates is undertaken. This is followed by an analysis of the more compositionally complex nano-size (<10 nm) precipitates.

### 4.1.1. Composition of Precipitates >90 nm in Size

As shown previously (Figure 4), precipitates > 90 nm in size are composed of Ti and Nb. The number frequency plot (Figure 7) indicates that the majority of the precipitates have a Ti composition of either 80–90 at. % or 30–40 at. %. For the Group 2 precipitates, the predicted average Ti composition of 87 at. % (Table 2) agrees well with the EDX analysis, both in terms of the Ti composition and the relatively high weight fraction of this group (31.7 wt. %) in the extracted residue. In addition, the predicted lattice parameter (Table 2) for the Group 2 precipitates ($Ti_{0.87}Nb_{0.13}N$) was 4.28 Å. This value corresponds reasonably well with the lattice parameter for a $Ti_xNb_{1-x}N$ ($0 \leq x \leq 1$) precipitate (4.26 Å) of this composition if a linear change in the lattice parameter, in going from TiN (4.24 Å) to NbN (4.40 Å), is assumed.

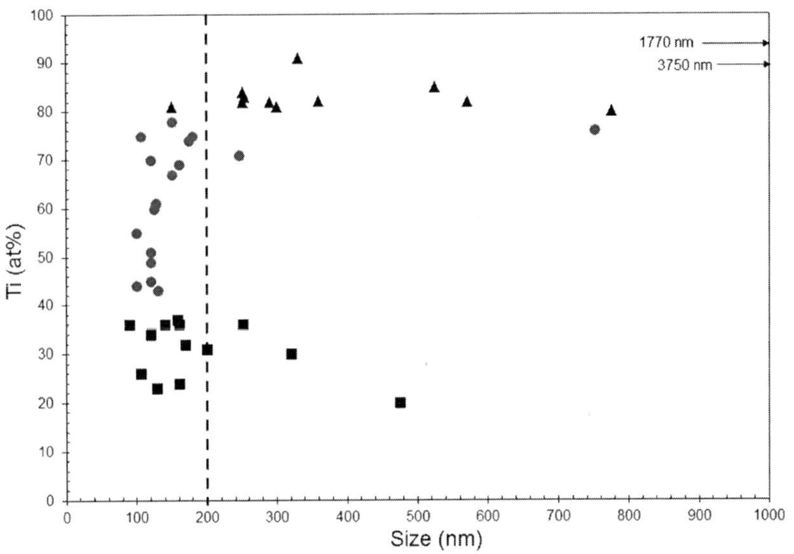


**Figure 6.** Ti composition *vs.* precipitate size, measured by EDX analysis, for the Group 1–3 precipitates.

QXRD predicts TiN (Group 1) precipitates are present, however, a TiN precipitate was not observed in the EDX analysis. This may be attributed to the relatively small number of precipitates (63) sampled.

## 4.1.2. Composition of Precipitates <10 nm in Size

The composition predicted by the Rietveld analysis for the Group 4 nano-size precipitates is $(Nb_{0.50}Ti_{0.32}M_{0.18})(C_{0.59}N_{0.41})$ (Table 2). The predicted Rietveld compositions (solid and dashed horizontal lines) are compared with the measured EDX analysis compositions for both Nb and Mo in Figure 8 and for both Ti and V in Figure 9. For both figures, the measured composition shows some deviation from the QXRD predicted compositions. In particular, the predicted level of V is zero even though EDX analysis confirmed its presence. It is believed that the amount of V present in the precipitates (<5 at. %) may be below the sensitivity of the quantitative XRD analysis.

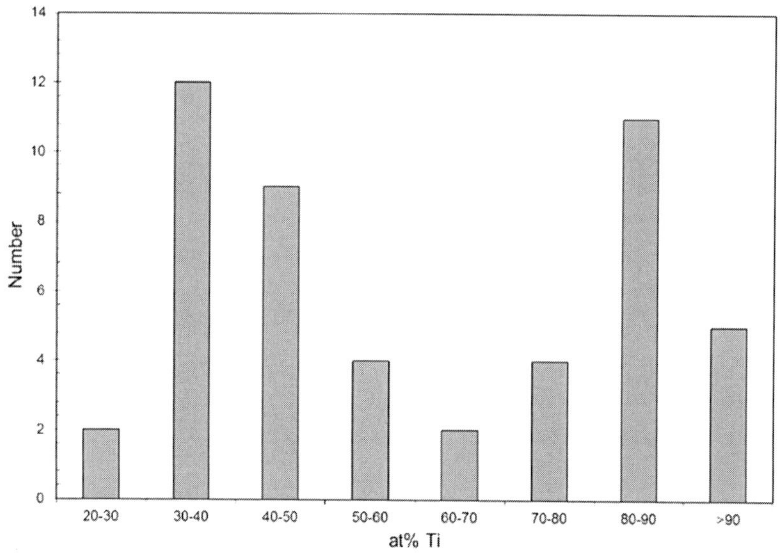


**Figure 7.** Number of precipitates, in Groups 2 and 3, as a function of measured Ti composition.

The deviations observed in Figure 8 and Figure 9 may be attributed to the relative difference in atomic scattering factors used in the QXRD intensity calculations (Equation (2)) for each atom type. The structure factors ($F$) are nearly identical for both Nb and Mo, since they have similar atomic scattering factors (they are adjacent to one other in the periodic table); 41 for Nb and 42 for Mo at $\sin\theta/\lambda = 0$. In an analogous manner, both Ti and V exhibit similar (though significantly smaller) atomic scattering factors; 22 for Ti and 23 for V at $\sin\theta/\lambda = 0$. The small difference in atomic scattering between each pair of atoms may be indistinguishable within the QXRD calculations. Figure 10 compares the total QRDX predicted values for both (Nb + Mo) composition and (Ti +

V) composition with the measured combined composition values. For both cases, there is better agreement between the measured composition values and the QXRD predicted values. These results indicate a limitation for using QXRD to distinguish between similar atoms in a precipitate crystal structure.

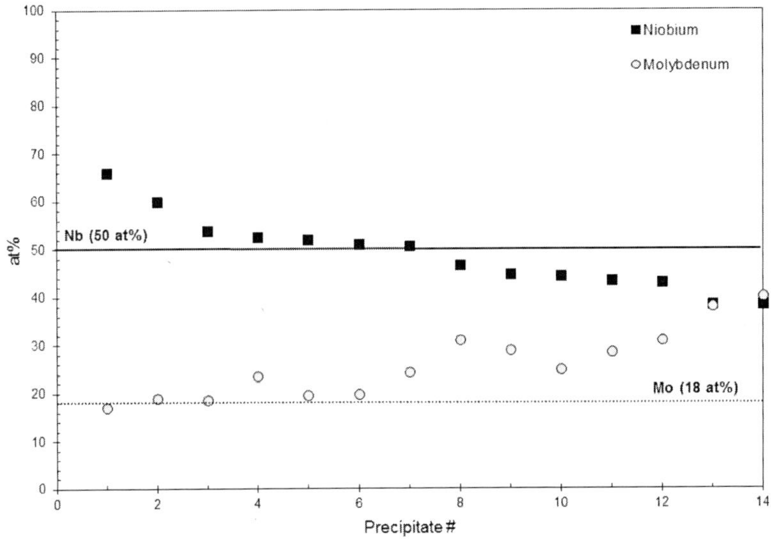


**Figure 8.** Measured Nb (squares) and Mo (circles) compositions *vs.* QXRD predicted compositions (solid line for Nb and dotted line for Mo).

In addition to the levels of the microalloying elements present, a relatively balanced amount of C and N ($C_{0.59}$ *vs.* $N_{0.41}$) is observed in the nano-precipitates (Table 2). This suggests that a sufficient amount of both C and N was still available in solution during the nano-precipitation event. However, the relative amount of C/N predicted by QXRD has not been independently confirmed.

## 4.2. Size ($L_{vol}$) Analysis

A comparison between the predicted and measured size of the large (>90 nm) precipitates has been undertaken. This is followed by a comparison of the predicted size of the nano-size (<10 nm) precipitates with SANS precipitate size results measured by the authors in previous work [10].

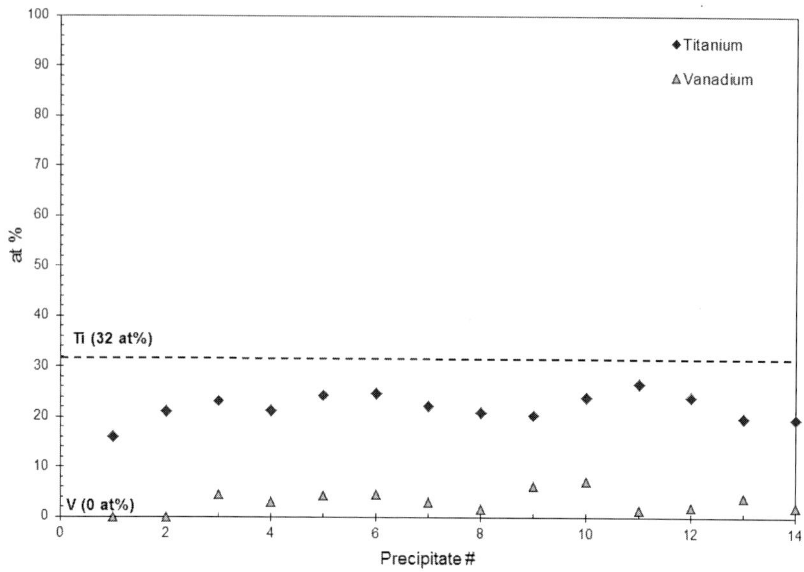


**Figure 9.** Measured Ti (diamonds) and V (triangles) compositions *vs.* QXRD predicted compositions (dashed line for Ti and 0 at % for V).

### 4.2.1. Size of Precipitates >90 nm

The wide variation in both composition (Figure 6) and size (Figure 7) for Group 1 to 3 precipitates is believed to have a strong effect on both the size ($L_{vol}$) and strain ($\varepsilon_o$) values predicted using QXRD.

The domain sizes ($L_{vol}$) predicted by quantitative XRD analysis (Table 2) for Group 1 to 3 precipitates (70 nm, 115 nm, and 40.3 nm) are smaller than the minimum size range reported in Figure 7. This contradiction between the size predicted by quantitative XRD and the size directly measured is likely related to the compositional complexity of the precipitates present in the Grade 100 steel. As discussed earlier, the predicted values of domain sizes (and microstrain) are based only on the broadening of the diffraction peak. However, artificially broadening of a diffraction peak can also arise from slight variations in lattice parameter (*i.e.*, a change in d spacing) due to a variation in the relative amounts of Ti and Nb in the Group 1–3 precipitates. Alternatively, XRD broadening may arise from a polycrystalline precipitate structure (*i.e.*, the domain size is a measure of the grain size within the precipitate and not the total precipitate size). However, selected area diffraction (SAD) analysis of the relatively large microalloyed steel precipitates [1] indicates the precipitates are single crystals; as such, the domain and precipitate size are equivalent.

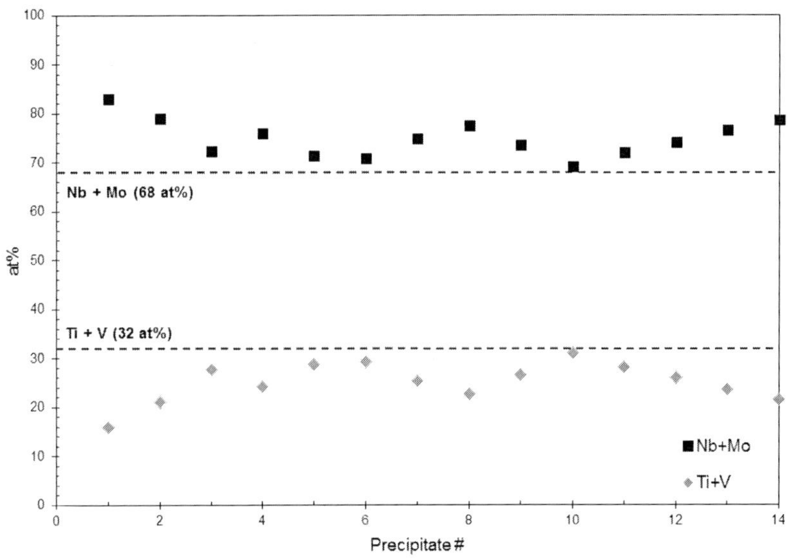


**Figure 10.** Measured (Nb + Mo) composition (squares) and measured composition (Ti + V) (diamonds) *vs.* predicted combined compositions (dashed lines).

## 4.2.2. Size of the Group 4 Precipitates

The average domain size ($L_{vol}$) predicted for the ($Nb_xTi_yMo_{1-x-y}$)(CN) precipitates ($0 \leq x, y \leq 1$) (Table 2) of 4.7 nm is in good agreement with the average size obtained using SANS (5.5 nm) [10] and TEM (3.9 nm) [28]. SAD analysis of the small size precipitates indicates that they are single crystals; hence, the value of $L_{vol}$ is equivalent to the precipitate size. It should be noted that the SANS and TEM data were originally obtained in the form of a size distribution. The TEM precipitate size (3.9

nm) corresponds to the size at which the maximum number of precipitates was observed. Similarly, the SANS analysis showed that the nano-size consisted of two overlapping log normal distributions with mean values of 3.4 nm and 6.6 nm [10]. A weighted average of each distribution was used to obtain the overall mean size of 5.5 nm from the SANS analysis. The effect of a multimodal distribution on the peak broadening effect in the Rietveld analysis is not known.

## 5. CONCLUSIONS

1) Quantitative X-ray diffraction (QXRD) via the Rietveld method can be used to quantify both the average size and composition of nano-size precipitates in a microalloyed steel. The average predicted $L_{vol}$ size of 4.7 nm corresponds well with precipitate size analysis conducted using transmission electron microscopy (TEM) (3.9 nm) and small angle neutron scattering (SANS) (5.5 nm). The predicted composition of the nano-size precipitates $((Nb_{0.50}Ti_{0.32}M_{0.18})(C_{0.59}N_{0.41}))$ also compares favorably with the individual compositions measured by energy dispersive X-ray (EDX) analysis.

2) The variation in the Nb and Ti composition (as measured with EDX) of the relative large (>90 nm) precipitates present in the Grade 100 steel limited the microstructure information that could be obtained from QXRD. In particular, the $L_{vol}$ predicted by quantitative XRD was inconsistent with the individual precipitate sizes physically measured.

3) QXRD can be used to quantify the effect of TCMP processing on precipitation in a microalloyed steel, as long as the limitations are adequately accounted for.

## ACKNOWLEDGMENTS

The authors would like to thank Laurie Collins for his input and EVRAZ N.A. Inc., Enbridge, TCPL, Alliance Pipelines and UT Quality for financial support.

## AUTHOR CONTRIBUTIONS

Lu completed her Ph.D. on precipitate characterization in microalloyed steels. Wiskel and Omotoso contributed to the QXRD section and Ivey and Henein contributed to the EDX/TEM section.

## REFERENCES

1. Sharma, U. Microstructural Characterization of Microalloyed Linepipe Steels. M.Sc. Thesis, University of Alberta, Edmonton, Canada, 1999.

2. Akhlaghi, S.; Ivey, D.G. Precipitation behaviour of a grade 100 structural steel. *Can. Metall. Q.* **2002**, *41*, 111–119.

3. Lu, J. Quantitative microstructural characterization of microalloyed steels. Ph.D. Thesis, University of Alberta, Edmonton, Canada, 2009.

4. Elwazri, A.M.; Varano, R.; Siciliano, F.; Bai, D.; Yue, S. Characterization of precipitation of niobium carbide using carbon extraction replicas and thin foils by FESEM. *Mater. Sci. Technol.* **2006**, *22*, 537–541.

5. Nagarajan, V.; Palmiere, E.J.; Sellars, C.M. New approach for modelling strain induced precipitation of Nb(C,N) in HSLA steels during multipass hot deformation in austenite. *Mater. Sci. Technol.* **2009**, *25*, 1168–1174.

6. Charleux, M.; Poole, W.J.; Militzer, M.; Deschamps, A. Precipitation behaviour and its effect on strengthening of an HSLA-Nb/Ti steel. *Metall. Mater. Trans. A* **2001**, *32A*, 1635–1647.

7. Shanmugam, S.; Ramisetti, N.K.; Misra, R.D.K.; Hartmann, J.; Jansto, S.G. Microstructure and high strength-toughness combination of a new 700 MPa Nb-microalloyed pipeline steel. *Mater. Sci. Eng. A* **2008**, *A478*, 26–37.

8. Lee, W.; Hong, S.G.; Park, C.G.; Park, S.H. Carbide precipitation and high-temperature strength of hot-rolled high-strength, low-alloy steels containing Nb and Mo. *Metall. Mater.Trans. A* **2002**, *33A*, 1689–1698.

9. Fatehi, A.; Calvo, J.; Elwazri, A.M.; Yue, S. Strengthening of HSLA steels by cool deformation. *Mater. Sci. Eng. A* **2010**,*A27*, 4233–4240.

10. Wiskel, J.B.; Ivey, D.G.; Henein, H. The effects of finish rolling temperature and cooling interrupt conditions on precipitation in microalloyed steels using small angle neutron scattering. *Metall. Trans. A* **2008**, *39B*, 116–124.

11. Hua, M.; Garcia, C.I.; Deardo, A.J. Precipitation behaviour in ultra-low-carbon steels containing titanium and niobium. *Metall. Trans. A* **1997**, *28A*, 1769–1780.

12. Deardo, A.J. Niobium in modern steels. *Int. Met. Rev.* **2003**, *48*, 371–402.

13. Cheary, R.W.; Ma-Sorrell, Y. Quantitative phase analysis by X-ray diffraction of martensite and austenite in strongly oriented orthodontic stainless steel wires. *J. Mater. Sci.* **2000**, *35*, 1105–1113.

14. Martiez-Blanco, D.; Gorria, P.; Blanco, J.A.; Perez, M.J.; Campo, J. Analysis of the diffraction-line broadening on nanostructured Fe: Size-strain effects induced by milling and heating. *J. Phys. Cond. Mater.* **2008**, *2*, 1–10.

15. Balzar, D.; Audebrand, N.; Daymond, M.R.; Fitch, A.; Hewat, A.; Langford, J.I.; le Bail, A.; Loue, D.; Masson, O.; McCowan, C.N.; *et al.* Size-strain line-broadening analysis of the ceria round-robin sample. *J. Appl. Crystallogr.* **2004**,*37*, 911–924.

16. Pourghahramani, P.; Forssberg, E. The characterization of structural changes in hematite ground in a confined particle bed using Rietveld analysis. *Int. J. Miner. Process.* **2007**, *83*, 47–59.

17. Langfordy, J.I.; Louer, D. Powder diffraction. *Rep. Prog. Phys.* **1996**, *59*, 131–234.

18. Young, R.A. Introduction to the Rietveld Method. In *The Rietveld Method*; Young, R.A., Ed.; Oxford University Press: Oxford, UK, 1993; pp. 1–30.

19. Snyder, R.L. Analytical profile fitting of X-ray powder diffraction profiles. In *The Rietveld Method*; Young, R.A., Ed.; Oxford University Press: Oxford, UK, 1993; pp. 112–131.

20. Delhez, R.D.; de Keijser, T.H.; Langford, J.I.; Louer, D.; Mittemeijer, E.J.; Sonneveld, E.J. Crystal imperfection broadening and peak shape in the Rietveld method. In *The Rietveld Method*; Young, R.A., Ed.; Oxford University Press: Oxford, UK, 1993; pp. 132–166. [Google Scholar]

21. Kuzel, R.; Holy, V.; Cernansky, M.; Kubena, J.; Simek, D.; Kub, J. Study of submicrocrystalline materials by diffuse scattering in transmitted waves. In *Diffraction Analysis of the Microstructure of Material*; Mittemeijer, E.J., Scardi, P., Eds.; Springer: New York, NY, USA, 1984; pp. 229–245.

22. Balzar, D.; Popa, N.C. Crystallite size and residual strain/stress modelling in Rietveld refinement. In *Diffraction Analysis of the Microstructure of Material*; Mittemeijer, E.J., Scardi, P., Eds.; Springer: New York, NY, USA, 1984; pp. 125–145.

23. Coelho, A. Topas Academic Version 4.1; software for analysis of powder diffraction data, Coelho software, Brisbane, Australia, 2007.

24. Scardi, P; Leoni, M. Whole profile pattern modelling: theory and applications. In *Diffraction Analysis of the Microstructure of Material*; Mittemeijer, E.J., Scardi, P., Eds.; Springer: New York, NY, USA, 1984; pp. 51–91.

25. Balzar, D. Voigt function model in diffraction-line broadening analysis. In *Microstructure Analysis from Diffraction*; Snyder, R.L., Bunge, H.J., Fiala, J., Eds.; Oxford University Press: Oxford, UK, 1999; pp. 94–124.

26. He, J.; Laverna, E.J.; Mateczyk, D.; Bampton, C.; Schoenung, J.M. Quantitative analysis of grain size in bimodal powders by X-ray diffraction and transmission electron microscopy. *J. Mater. Sci.* **2004**, *39*, 6957–6964.

27. Vives, S.; Gaffet, E.; Meunier, C. X-ray diffraction line profile analysis of iron ball milled powders. *Mater. Sci. Eng.***2004**, *A366*, 229–238.

28. Poorhaydari-Anaraki, K. Microstructure and Property Examination of the Weld HAZ in Grade 100 Microalloyed Steel. Ph.D. Thesis, University of Alberta, Edmonton, Canada, 2005.

6

# ARTIFICIAL NEURAL NETWORKS INVESTIGATION OF INDENTATION FORCE EFFECTS ON NANO- AND MICROHARDNESS OF DUAL PHASE STEELS

**A. Fotovati,[1] J. Kadkhodapour,[2] and S. Schmauder[1]**

[1]Institute for Materials Testing, Materials Science and Strength of Materials, University of Stuttgart, 70569 Stuttgart, Germany

[2]Mechanical Engineering Department, Shahid Rajaee Teacher Training University, Tehran 16758-136, Iran

## ABSTRACT

Nanoindentation test results on different grain sizes of dual phase (DP) steels are used to train artificial neural networks (ANNs). With selection of ferrite and martensite grain size, martensite volume fraction (MVF), and indentation force as input and microhardness, ferrite, and martensite nanohardness as outputs, six different ANNs are trained according to normalized datasets to predict hardness and their tolerances. A graphical user interface (GUI) is developed for a better investigation of the trained ANN prediction. The response of the ANN is analyzed in five case studies. In each case the variation of two input parameters on the output is analyzed when the other input parameters are kept constant. Reliable and reasonable results of ANN predictions are achieved in each case.

# 1. INTRODUCTION

During the last decades, very strong improvements in steel industries lead to produce advanced high strength steels (AHSS) which have a strong competition with low density metals in automobile industries to passenger safety improvement, vehicle performance, and fuel economy [1]. Among these alloys, DP steels are very interesting for light weight constructions because they have a high total elongation (TE) in their stress-strain curve which leads to a good formability. Another advantage of this material is its low yield strength, high hardening ratio, and the absence of discontinuous yielding. DP sheets are well suited for forming of automotive structural and crash relevant parts such as cross members, longitudinal beams, and reinforcements.

DP steels are based on a microstructure consisting of a ferritic matrix with a second phase of martensite. The size, morphology, and volume fraction of the martensite phase depend on the production process and heat treatment procedure, especially the intercritical annealing temperature and the heating and cooling rate. The variation of the martensite morphology ranges from single martensite islands which are arranged in between the ferrite grains up to a more layer-wise arrangement of the martensite with a distinct orientation of the layer. There may also exist a small amount of bainite and retained austenite in the microstructure. The most important features influencing the mechanical properties of the dual phase microstructure are shape, size, volume fraction and distribution of ferrite and martensite, carbon content of martensite, and volume fraction of retained austenite. Independently of the specific microstructure of dual phase steels, the following basic principle exists: the hard and more brittle martensite phase is distributed in a soft and more ductile ferrite phase to produce a composite mixture.

On nanoscale, extensive investigations were carried out to determine local hardening in the structure of DP steels in the past [2]. In Figures 1(a) and 1(b), local hardening in a typical ferrite grain in a DP steel is illustrated. Investigations at different locations show that the extent and rate of hardening vary in each grain, but the general trend is the same within all ferrite grains. In all cases, hardening can be observed near phase boundaries. Local hardening in this area is related to the presence of geometrically necessary dislocations (GNDs). Measurements indicate that GNDs close to the martensite boundaries are around one order of magnitude higher than GND densities inside the ferrite grains.

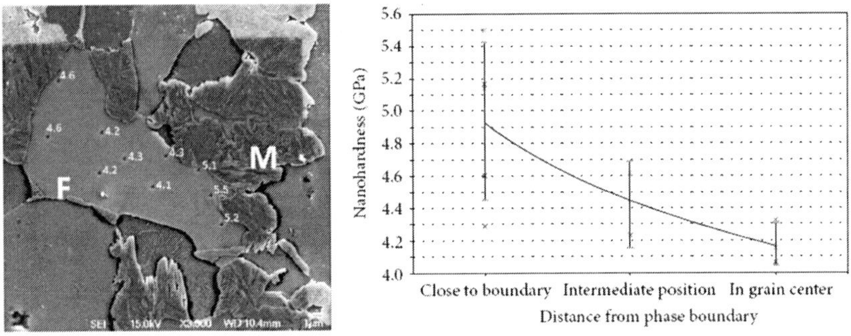


(a) (b)

**Figure 1.** Local hardening near the grain boundary in DP steel; (a) nanohardness measurements in ferrite grain, (b) effect of phase boundary on nanohardness [3].

Among a wide range of investigations on DP steels [2–9], the effect of nanoindentation forces on nano- and microhardness is not clearly defined. The only work that clearly shows the force effects on hardness in various grain sizes in DP steel is done by Delincé et al. [10] which is the base of creating the datasets of this paper. Figure 2 shows exactly the effect of indentation force on the hardness in the case of a Berkovich tip. In point 1 the indentation force is high enough to show the hardness of the pure ferrite phase only. In point 2 the effect of grain boundary strengthening is contained in the results of nanohardness investigation. In point 3 the effect of martensite phase is added and in point 4 the indentation force is so high and shows an average hardness of the second phase, grain boundaries, and sizes. Depending on the indentation depth (), the results can be categorized in nano- or microhardness.

To find out exactly the influence of these parameters on hardness, ANN is the best choice. Krajewski and Nowacki [11] used ANN to model the relationship of chemical composition and mechanical properties of DP steels. Bahrami et al. [12] developed ANN to predict mechanical properties of DP steels. Haque and Sudhakar [13] investigated corrosion-fatigue crack growth of DP steels using ANN. Other authors [14–17] also used ANN in low carbon steels for the prediction of its properties. All of them showed that a well-trained ANN is a powerful and reliable tool for predicting new situations based on datasets used in a training procedure.

Developing a procedure that could be well organized and predict the effect of indentation force on nano- and microhardness for any new situation is essentially needed for future researches. For achieving this goal, ANN is powerful tool for learning the influence of force on hardness and estimating other interested situations according to the new desired problems.

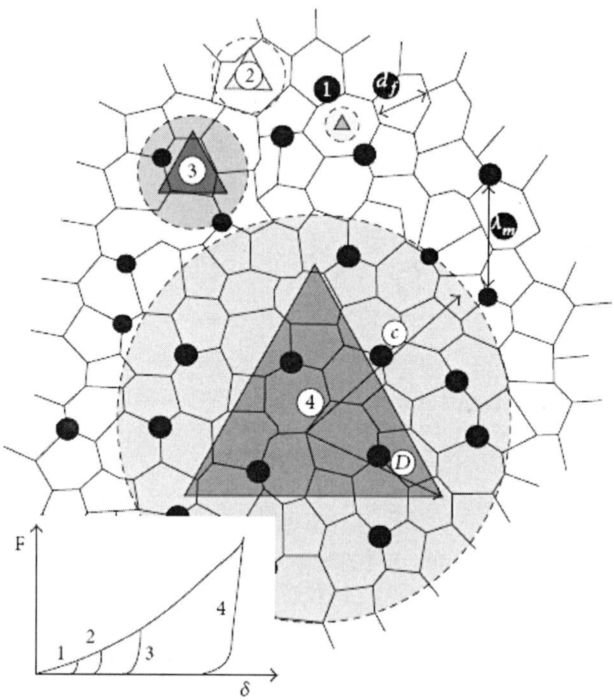


**Figure 2.** Schematic description of the interactions between the indentation plastic zone and the microstructure as a function of the indentation force [1].

## 2. ANN MODEL BUILD-UP

For making a good prediction of nano- and microhardness of DP steels, a back-propagation ANN is used. This kind of network is frequently used for function approximation, where there is a relation between input parameters, here ferrite and martensite grain sizes, MVF and indentation force, and output parameters such as nano- and microhardness and their tolerances. For training such networks, there is a training algorithm, for adjusting the weights and bias of each neuron in such a way that the mean square error (MSE) of predicted and real target is minimized in each iteration. A Levenberg-Marquardt algorithm is used for this research due to its fast convergence [18]. MSE is calculated according to the following formula [12]:

$$MSE = \frac{1}{QN_O}\sum_{m=1}^{Q}\sum_{n=1}^{N_O}\left[d_n(m) - y_n(m)\right]^2, \tag{1}$$

where is the number of output nodes, is the number of training sets, is the target, and is the network prediction calculated according to the following formula [12]:

$$y_i = f\left(\sum_{j=1}^{n} w_{ij} + b\right). \tag{2}$$

Here, is the weight of each neuron in the related layer and is the bias, while is the activation function. This function is the most important part of the learning procedure and in this research the tangent sigmoid function is used for hidden layers and a linear function is used for the output layer. As the output of the tangent sigmoid function is a number in the range of [−1 +1], it is a good practice to normalize the input and target parameters to this range to help the network learn much quicker and more reliable since the importance of large numbers in comparison with decimal digits is diminished. Figure 3 shows a typical arrangement of hidden and output layers of an ANN and connection of neurons through weights and biases. This configuration is generally called the topology of the ANN.

The training is started with a practical dataset of reference [10] with programming in the Neural Network Toolbox of MATLAB [18]. The summary of the training set for nanohardness of ferrite and martensite and their tolerances is presented in Table 1. Moreover, Table 2 shows the training set for the microhardness prediction. Although there is a possibility to train a multilayer ANN with more than one output layer, it is not really clear how the output of the trained networks is reliable. For this reason, six different ANNs are trained according to the normalized training parameters of Tables 1 and 2. During the training procedure, three groups of data are randomly selected from the whole datasets: training, validation, and test. Training sets that consist of 70% of datasets are the main series for training the networks. Validation and test sets are used to evaluate only the performance of the ANN training procedure to prevent the network from overfitting or memorizing. Each of these sets is selected randomly from 15% of datasets, respectively. In this way MSE of three sets is calculated after each iteration (epoch) and then the deviation between them is measured. If this amount is in the range of the training parameters the next epoch is started; otherwise, the training is stopped. As these three sets are selected randomly at the start of training, it is important to analyze the performance of the trained network with regression coefficient that is the best linear fit between the target and the predicted values to achieve the goal of  for the whole datasets.

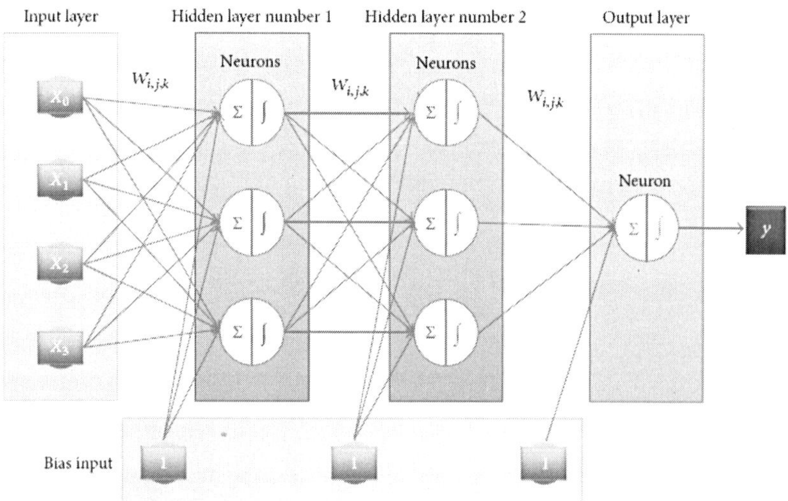


**Figure 3.** Typical arrangement of ANN layers and their connections by weight and biases.

The number of hidden layers is selected according to the best regression coefficient from a large attempt of training. The summary of networks topology (number of hidden and output layers), regression coefficient (), and MSE of three training sets are summarized in Table 3. When there are simple relations between input and output sets, a better  is achieved with a lower number of neurons in the hidden layers and when the complexity of data increased, the need of a higher number of neurons and hidden layers is desirable.

**Table 1.** Range of 21 training parameters for nanohardness.

| | Training parameters | Min | Max | Avr. | Std. Dev. |
|---|---|---|---|---|---|
| Input | F grain size (μm) | 0.7 | 3.9 | 1.87 | 1.48 |
| | M grain size (μm) | 0.82 | 1.4 | 1.03 | 0.27 |
| | MVF (%) | 22 | 34 | 27.33 | 5.11 |
| | Force (mN) | 0.25 | 12 | 3.96 | 4.23 |
| Target | F nanohardness (Gpa) | 2.31 | 4.2 | 2.95 | 0.49 |
| | F nanohardness tolerance (Gpa) | 0.05 | 0.72 | 0.29 | 0.19 |
| | M nanohardness (GPa) | 4.56 | 7.62 | 5.86 | 0.93 |
| | M nanohardness tolerance (GPa) | 0.25 | 5.75 | 1.58 | 1.4 |

**Table 2.** Range of 18 training parameters for microhardness.

| | Training parameters | Min | Max | Avr. | Std. Dev. |
|---|---|---|---|---|---|
| Input | F grain size (μm) | 0.7 | 3.9 | 2.08 | 1.5 |
| | M grain size (μm) | 0.82 | 1.4 | 1.05 | 0.28 |
| | MVF (%) | 22 | 34 | 25.78 | 4.22 |
| | Force (mN) | 20 | 20000 | 4629 | 7279 |
| Target | Microhardness (Gpa) | 2.53 | 4.02 | 3.07 | 0.41 |
| | Microhardness tolerance (Gpa) | 0.05 | 0.77 | 0.27 | 0.22 |

# 3. RESULTS AND DISCUSSION

Since there are six trained ANNs and working with them without a computer is too difficult, a GUI is developed in MATLAB. Figure 4 shows the screenshot of this package. There are three main areas designed for working: input parameters, predicted values, and plot area. As any parameter is defined in the input area, the software normalizes each input according to the rule of training and calculates automatically the nano- and microhardness and their tolerances according to the trained ANN and shows them in the related area. There is also a possibility to examine the effect of input parameters on the micro- and nanohardnesses and their tolerances in the plot area. This software is free to download from the IMWF website (http://www.imwf.uni-stuttgart.de/forschung/downloads).

According to previous works [19–21], a comparison between datasets used for training and validation of the ANN and the predicted results from ANN is illustrated in Figure 5. In Figure 5(a), the experimental data [10] for ferrite nanohardness is used. These data are presented for three DP steels: coarse grain (CG), fine grain (FG), and very fine grain (VFG). The specific detail of grain size and MVF for each category is presented in Table 4. The predicted results from the trained ANN according to fixed parameters of Table 4 are calculated and then the difference from real data drawn in this figure for three types of DP steels as error bars. As there is a smooth trend in experimental data, there is no difficulty for the ANN to learn the relations between input and output parameters and as indicated in Table 3 the regression coefficient of 0.99 is achieved. Also a close prediction to real data is watchable in this figure for the ANN output.

Figure 5(b) shows the difference between experimental data used for training the ANN and predicted data for martensite nanohardness. As it is observable, there is a fluctuation in datasets and it is a difficult situation for the ANN to learn the rule. As it is reported in Table 3 the regression coefficient for this ANN is 0.89 with a larger hidden layer than the previous ANN. According to this complexity of trained data the response of the ANN is acceptable.

In Figure 5(c) there is a shortage of experimental data especially in lower indentation forces. This will affect training of the ANN and makes it difficult to learn the rule by the ANN. According to Table 3 with a 30-30-1 ANN topology the regression coefficient of 0.93 is archived which is acceptable for this work. Although there is a good reached in this ANN, there is some super estimation or exaggerated results predicted by this network that is shown in Sections 3.3 and 3.4. Although there is no data presented to the ANN during training for VFG, the predicted data for microhardness of this steel is reasonable and it shows the power of the ANN prediction in new situations.

**Table 3.** Trained ANN specifications according to normalized datasets in range of [−1  +1].

| Trained ANNs | Topology | *R* | MSE | MSE *v* | MSE *t* |
|---|---|---|---|---|---|
| F nanohardness (Gpa) | 20-10-1 | 0.99 | 0.0041 | 0.0015 | 0.0173 |
| M nanohardness (Gpa) | 40-20-1 | 0.89 | 0.0239 | 0.0925 | 0.3209 |
| Microhardness (Gpa) | 30-30-1 | 0.93 | 0.0370 | 0.0429 | 0.0631 |
| F nanohardness tolerance (Gpa) | 20-10-1 | 0.98 | 0.0019 | 0.0338 | 0.0484 |
| M nanohardness tolerance (Gpa) | 35-20-1 | 0.93 | 0.0043 | 0.1965 | 0.0771 |
| Microhardness tolerance (Gpa) | 40-20-1 | 0.84 | 0.1014 | 0.2065 | 0.0227 |

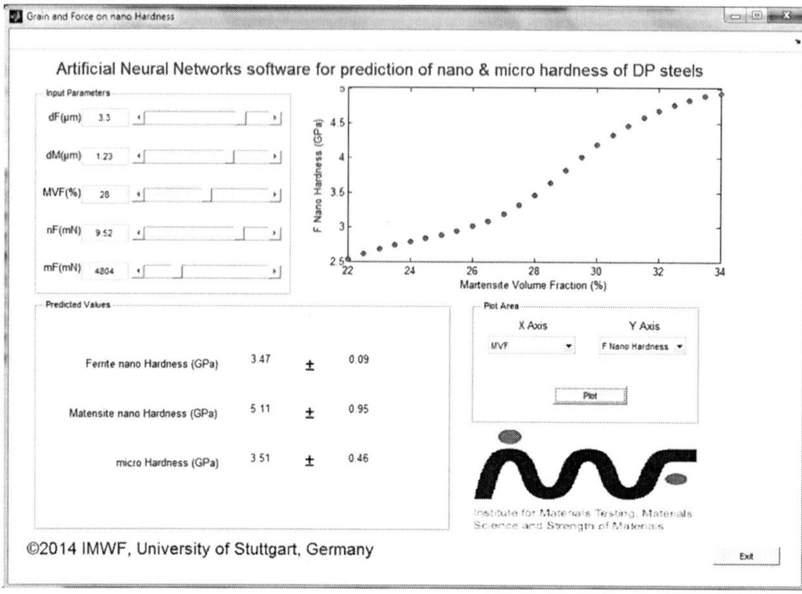


**Figure 4.** Screenshot of ANN GUI developed for this research.

**Table 4.** Ferrite and martensite grain size and MVF for three DP steels [1].

| | CG | FG | VFG |
|---|---|---|---|
| F grain size ($\mu$m) | 3.9 | 1.0 | 0.7 |
| M grain size ($\mu$m) | 1.4 | 0.82 | 0.87 |
| MVF (%) | 22 | 26 | 34 |

For a better understanding of the trained ANN prediction and its response to a variation in the input data, five case studies on hardness prediction are defined according to Table 5.

## 3.1. Effects of Ferrite and Martensite Grain Sizes

In this case the effect of ferrite and martensite grain sizes on nano- and microhardness is investigated when MVF and indentation force are constant. Figure 6(a) shows these effects on the ferrite nanohardness. According to Table 5 the nanoforce is selected as 3 mN. With a look around in Figure 5(a), it is expectable that the ANN predicted the ferrite nanohardness with an error less than 0.05 GPa. There is a slight increase in the ferrite nanohardness with increasing of the ferrite grain size (although there is a first decrease in the ferrite nanohardness for larger martensite grain size and then an increase) and a reverse situation for martensite grain sizes with a higher amplitude is observed. This prediction is completely compatible with the role of carbon content in ferrite and martensite grains [10]. With this prediction the maximum ferrite nanohardness is achievable in fine martensite and coarse ferrite grains.

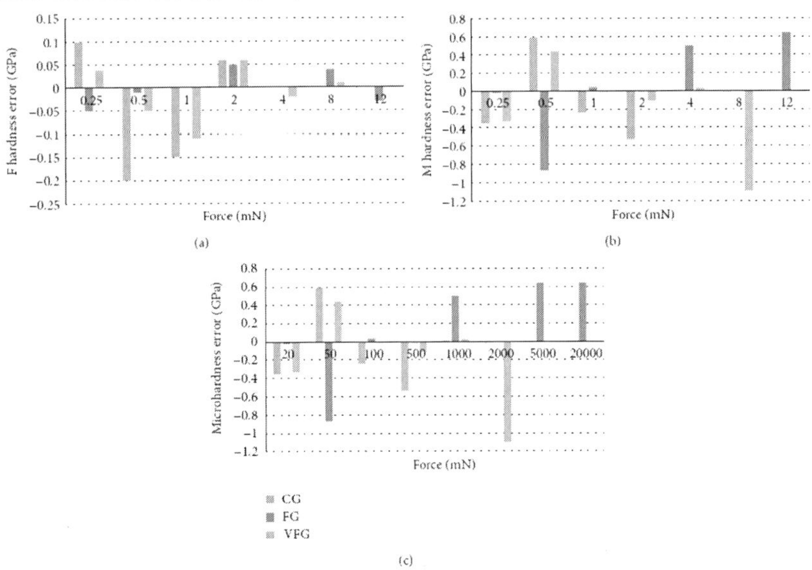


**Figure 5.** The error bar of trained ANN for real data that is used for training, for different types of grain size. These graphs show the effect of indentation force on (a) F (ferrite) nanohardness, (b) M (martensite) nanohardness, and (c) microhardness.

Figure 6(b) shows these effects on the martensite nanohardness. There exist more complex relations in the effects of grain size on martensite nanohardness. Indeed, the maximum martensite nanohardness is predictable in fine martensite and ferrite grain sizes.

**Table 5.** Input parameters range for trained ANN.

| Input parameters | Case 1 | Case 2 | Case 3 | Case 4 | Case 5 |
|---|---|---|---|---|---|
| F grain size (μm) | ~ | 3.2 | 2.2 | ~ | 1.1 |
| M grain size (μm) | ~ | ~ | 1.2 | 0.93 | ~ |
| MVF (%) | 26 | ~ | ~ | 24 | 26 |
| Nanoforce (mN) | 3 | 3 | ~ | ~ | ~ |
| Microforce (mN) | 18000 | 3000 | ~ | ~ | ~ |

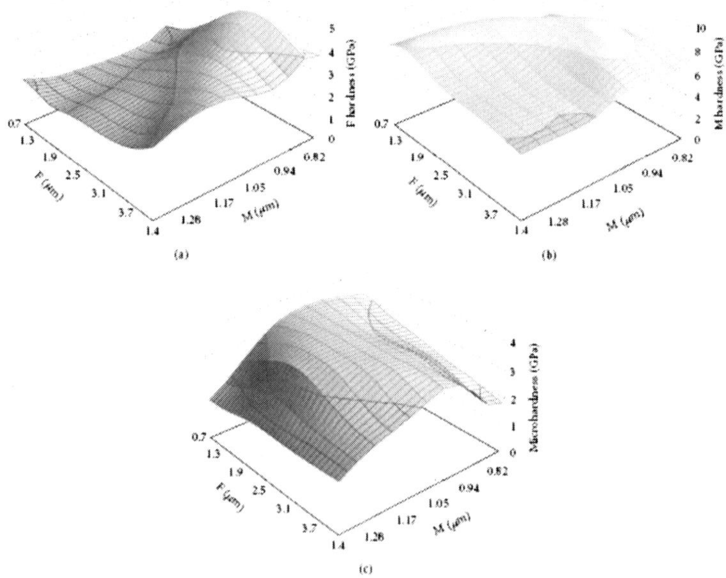


**Figure 6.** Effects of ferrite and martensite grain sizes on (a) F (ferrite) nanohardness, (b) M (martensite) nanohardness, and (c) microhardness.

The higher microhardness is achievable by the finer martensite grains according to Figure 6(c). This prediction confirms that the finer grains produce higher strength steels because of the increase of grain boundaries and an increase in the stacking faults. According to Hall-Petch effect, martensite produces more strength than ferrite when its grain is reduced. This phenomenon is according to the type of crystalline network of martensite in blocking of dislocations and hence increasing the hardness of DP steel. Further works about this matter will be done in future publications.

## 3.2. Effects of Martensite Grain Size and MVF

The effects of martensite grain size and volume fraction on ferrite nanohardness are illustrated in Figure 7(a). Increasing of MVF and decreasing of martensite grain size lead to higher ferrite nanohardness. It is due to the increase of the carbon content in the ferrite matrix. A similar effect is also predictable on martensite nanohardness in Figure 7(b), but there is an apex in MVF of around 30% at the minimum martensite grain size.

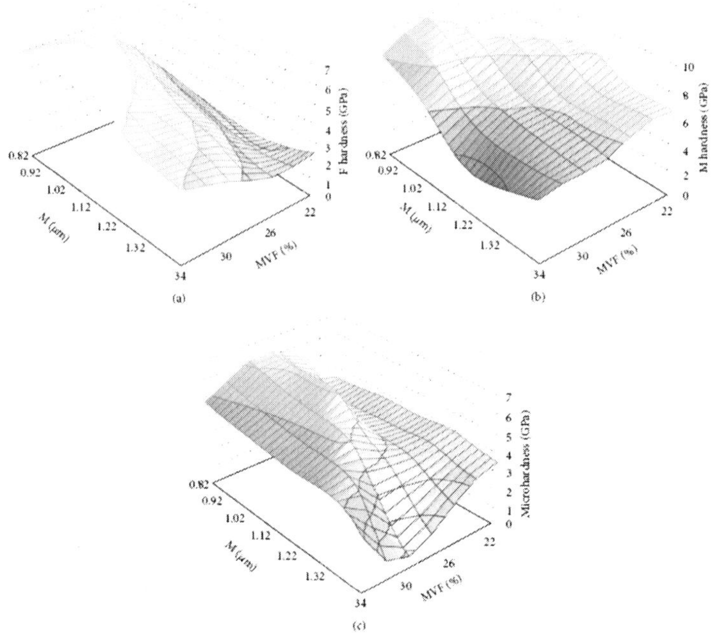


**Figure 7.** Effects of martensite grain size and volume fraction (MVF) on (a) F (ferrite) nanohardness, (b) M (martensite) nanohardness, and (c) microhardness.

A similar effect is detectable in Figure 7(c) where the maximum microhardness is predicted at lower martensite grain sizes and an MVF of 28%.

Although there are some differences in these figures, all of them show that an increase in the martensite grain size will decrease the hardness while the MVF has more effect in smaller martensite grain size.

## 3.3. Effects of MVF and Indentation Force

Figure 8(a) shows the effects of MVF and indentation force on ferrite nanohardness. As it is obvious, the indentation force has a slight reduction effect on ferrite nanohardness due to the increase of the indentation area and more grains affecting the result. Perhaps at low forces the result is related to single grain and at high forces it is averaged over some grains and second phase and hence a slight reduction of nanohardness is predictable through the ANN. For MVF a more increasing effect on ferrite nanohardness is calculated and it is completely reasonable due to the increase of the carbon content in the ferrite matrix. Indeed, the maximum hardness is predictable at higher MVF and lower indentation force.

The effect of these parameters on martensite nanohardness is illustrated in Figure 8(b). Almost the MVF increase will cause a hardness reduction in martensite but an increase in force has a complicated effect where for low MVF it has a downward and then an upward trend while for higher MVF it will reverse. However, the maximum amount of martensite nanohardness is predictable at 4 mN force and 22% of MVF.

Similar results are observable in Figure 8(c) for microhardness values. The maximum hardness is predictable at 10 N force and 22% MVF. One important point that is observable in this figure is the jump in the surface area in the position of higher MVF and lower indentation force. As during the training procedure, the data presented in this region by the ANN are not sufficient, so there is an overestimate from the prediction network. Also remember that the regression coefficient for this network is 0.93 and it means that the results are 93% reliable. Moreover, there is a shortage in experimental data as mentioned in the previous section that affected the ANN prediction.

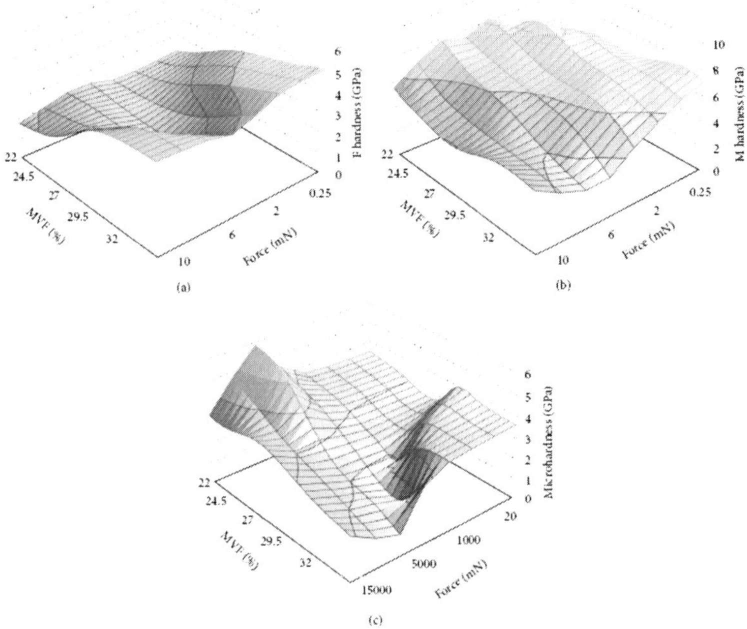


**Figure 8.** Effects of martensite volume fraction (MVF) and indentation force on (a) F (ferrite) nanohardness, (b) M (martensite) nanohardness, and (c) microhardness.

## 3.4. Effects of Ferrite Grain Size and Indentation Force

An increase in the ferrite grain size will increase the ferrite nanohardness at low indentation force (Figure 9(a)) while at higher forces this gradient becomes reverse. On the other hand, there is a decrease in ferrite nanohardness at higher forces as it is predictable by ANN. Meanwhile, steady changes from a positive slope to a negative one can be observed for indentation forces in such a way that the maximum point is located at 2.8 μm ferrite grain size and at 0.25 mN indentation force.

In Figure 9(b) there is a reverse prediction for martensite nanohardness in comparison with Figure 9(a). The maximum hardness is predicted at 2.4 μm ferrite grain size and for 4 mN force.

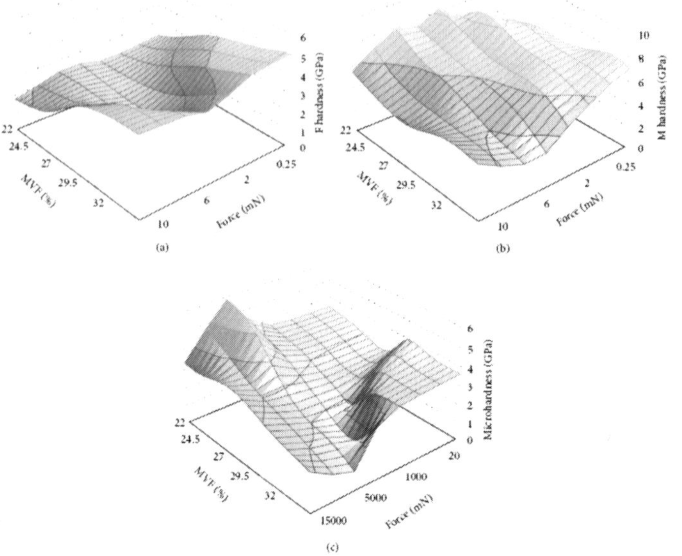


**Figure 9.** Effects of ferrite grain size and indentation force on (a) F (ferrite) nanohardness, (b) M (martensite) nanohardness, and (c) microhardness.

Figure 9(c) shows the effect of force and ferrite grain size on microhardness. For fine grains, there is no such force dependency in microhardness since the microstructure is a uniform mixture of ferrite and martensite. In case of increased ferrite grains, since martensite grain size is constant, a reverse effect in microhardness is predictable. The peak section in this figure is a super estimation by ANN and is not reasonable. Since for the lower indentation force there was no experimental data for training the ANN, prediction of the ANN is not correct for this section (according to Table 2, the maximum microhardness is 4.02).

## 3.5. Effects of Martensite Grain Size and Indentation Force

In the previous four cases, the effects of some parameters are investigated on hardness. Although there are more investigations, the effects of martensite grain size and indentation force on hardness are selected for the last case as illustrated in Figure 10. In Figure 10(a) increasing of martensite grain size leads to a decrcase of ferrite nanohardness over all ranges of indentation forces. The maximum hardness is predictable for small martensite grains as it is expected from the experimental observation that the finer grain size has the higher hardness.

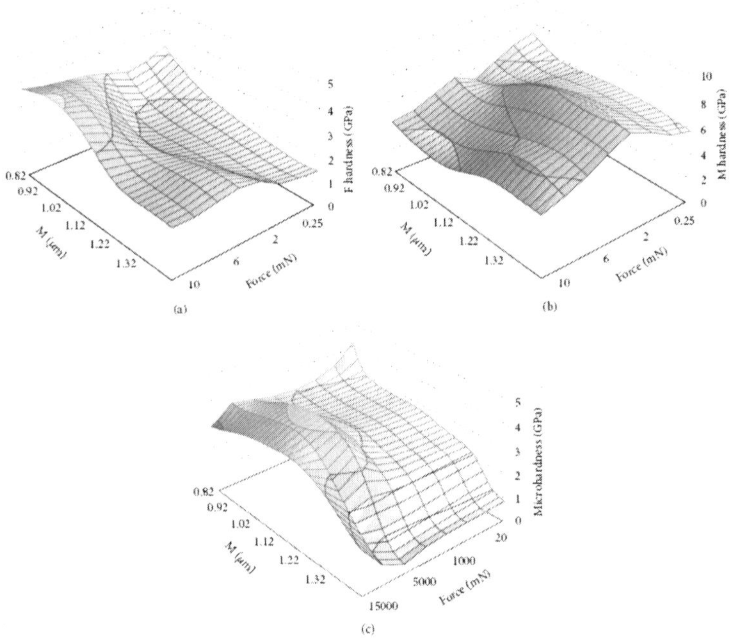


**Figure 10.** Effects of martensite grain size and indentation force on (a) F (ferrite) nanohardness, (b) M (martensite) nanohardness, and (c) microhardness.

Figure 10(b) shows that the martensite grain size does not have any serious effect on martensite nanohardness and only a force increase will reduce the hardness. This phenomenon is reasonable since at lower indentation forces there is no balance between applied load and stiffness of the martensite phase and hence more reliable results are achievable for higher indentation forces. This discussion is completely extractable from Figure 10(c)where at higher indentation force there are no severe changes in microhardness. The downward trend of this curve completely approves that the finer martensite grain structure possesses a higher microhardness and this matter is more severe in comparison with the effect of ferrite grain size on microhardness as shown in Figure9(c).

# 4. CONCLUSIONS

Six multilayer back-propagation ANNs are trained in this research according to a normalized data base of reference [1]. Three of them predicted the microhardness and ferrite and martensite nanohardness of DP steel and the others forecasted the tolerances of each hardness. A GUI is developed for ease of usage of ANN. Five case studies are selected to show the power and flexibility of trained ANN. The following results are predictable by trained ANN as investigated in five case studies.(1)Fine ferrite and martensite grain sizes have higher ferrite and martensite nano- and microhardness.(2)MVF increases ferrite nanohardness when martensite grain size is small. But, it decreases martensite nanohardness and microhardness in coarse grains.(3) Indentation force has a decreasing effect on hardness for large MVF.(4)Nano- and microhardness will decrease when ferrite grain size and indentation force increase.(5)In the same way, nano- and microhardness will decrease when martensite grain size and to some extent indentation force increase.

# REFERENCES

1. R. Kuziak, R. Kawalla, and S. Waengler, "Advanced high strength steels for automotive industry: a review," Archives of Civil and Mechanical Engineering, vol. 8, no. 2, pp. 103–117, 2008.

2. J. Kadkhodapour, S. Schmauder, D. Raabe, S. Ziaei-Rad, U. Weber, and M. Calcagnotto, "Experimental and numerical study on geometrically necessary dislocations and non-homogeneous mechanical properties of the ferrite phase in dual phase steels," Acta Materialia, vol. 59, no. 11, pp. 4387–4394, 2011.

3. E. Fereiduni and S. S. G. Banadkouki, "Improvement of mechanical properties in a dual-phase ferrite-martensite AISI4140 steel under

tough-strong ferrite formation," Materials and Design, vol. 56, pp. 232–240, 2014.

4. E. Fereiduni and S. S. Ghasemi Banadkouki, "Reliability/unreliability of mixture rule in a low alloy ferrite-martensite dual phase steel," Journal of Alloys and Compounds, vol. 577, pp. 351–359, 2013.

5. K. Hayashi, K. Miyata, F. Katsuki, T. Ishimoto, and T. Nakano, "Individual mechanical properties of ferrite and martensite in Fe-0.16 mass% C-1.0 mass% Si-1.5 mass% Mn steel," Journal of Alloys and Compounds, vol. 577, supplement 1, pp. S593–S596, 2013.

6. V. H. B. Hernandez, S. K. Panda, M. L. Kuntz, and Y. Zhou, "Nanoindentation and microstructure analysis of resistance spot welded dual phase steel," Materials Letters, vol. 64, no. 2, pp. 207–210, 2010.

7. M. D. Taylor, K. S. Choi, X. Sun et al., "Correlations between nanoindentation hardness and macroscopic mechanical properties in DP980 steels," Materials Science and Engineering A, vol. 597, pp. 431–439, 2014.

8. G. Rosenberg, I. Sinaiová, and Ľ. Juhar, "Effect of microstructure on mechanical properties of dual phase steels in the presence of stress concentrators," Materials Science and Engineering A, vol. 582, pp. 347–358, 2013.

9. E. Fereiduni and S. S. Ghasemi Banadkouki, "Ferrite hardening response in a low alloy ferrite-martensite dual phase steel," Journal of Alloys and Compounds, vol. 589, pp. 288–294, 2014.

10. M. Delincé, P. J. Jacques, and T. Pardoen, "Separation of size-dependent strengthening contributions in fine-grained Dual Phase steels by nanoindentation," Acta Materialia, vol. 54, no. 12, pp. 3395–3404, 2006.

11. S. Krajewski and J. Nowacki, "Dual-phase steels microstructure and properties consideration based on artificial intelligence techniques," Archives of Civil and Mechanical Engineering, vol. 14, no. 2, pp. 278–286, 2014.

12. A. Bahrami, S. H. M. Anijdan, and A. Ekrami, "Prediction of mechanical properties of DP steels using neural network model,"

Journal of Alloys and Compounds, vol. 392, no. 1-2, pp. 177–182, 2005.

13. M. E. Haque and K. V. Sudhakar, "Prediction of corrosion-fatigue behavior of DP steel through artificial neural network," International Journal of Fatigue, vol. 23, no. 1, pp. 1–4, 2001.

14. H. Monajati, D. Asefi, A. Parsapour, and S. Abbasi, "Analysis of the effects of processing parameters on mechanical properties and formability of cold rolled low carbon steel sheets using neural networks,"Computational Materials Science, vol. 49, no. 4, pp. 876–881, 2010.

15. K. Dehghani and A. Shafiei, "Predicting the bake hardenability of steels using neural network modeling,"Materials Letters, vol. 62, no. 2, pp. 173–178, 2008.

16. P. Saravanakumar, V. Jothimani, L. Sureshbabu, S. Ayyappan, D. Noorullah, and P. G. Venkatakrishnan, "Prediction of mechanical properties of low carbon steel in hot rolling process using artificial neural network model," Procedia Engineering, vol. 38, pp. 3418–3425, 2012.

17. K. Dehghani and A. Nekahi, "Artificial neural network to predict the effect of thermomechanical treatments on bake hardenability of low carbon steels," Materials and Design, vol. 31, no. 4, pp. 2224–2229, 2010.

18. M. Hudson, M. Hagan, and H. Demuth, Neural Network Toolbox User's Guide, Version 8.0.1, The MathWorks, 2013.

19. A. Mukherjee, S. Schmauder, and M. Rühle, "Artificial neural networks for the prediction of mechanical behavior of metal matrix composites," Acta Metallurgica Et Materialia, vol. 43, no. 11, pp. 4083–4091, 1995.

20. A. Fotovati and T. Goswami, "Prediction of elevated temperature fatigue crack growth rates in TI-6AL-4V alloy—neural network approach," Materials and Design, vol. 25, no. 7, pp. 547–554, 2004.

21. A. Moayyed, V. Fallah, and A. Fotovati, "Modeling of hot deformation behavior of equiaxed Ti-6Al-4V ELI in two phase ($\alpha$+ $\beta$) region using neural networks," in Proceedings of the 11th World Conference on Titanium, pp. 319–322, Kyoto, Japan, 2007.

7

# ALUMINIUM AA6061 MATRIX COMPOSITE REINFORCED WITH SPHERICAL ALUMINA PARTICLES PRODUCED BY INFILTRATION: PERSPECTIVE ON AEROSPACE APPLICATIONS

**Claudio Bacciarini[1,2] and Vincent Mathier[1]**

[1]Constellium Innovation Cells, EPFL-Innovation Park, Bâtiment E, 1015 Lausanne, Switzerland

[2]JetSolutions SA, Route de Montena 89, 1728 Rossens, Switzerland

## ABSTRACT

Metal matrix composites, based on AA6061 reinforced with 60 vol% spherical particles, were produced by gas pressure infiltration and characterized for hardness, impulse excitation modulus, tensile properties (at room temperature and at 250°C), and machining. It was experimentally demonstrated that the novel alumina powder used in the present work does not react with the liquid Mg-containing matrix during the infiltration process. The AA6061 matrix therefore retains its ability to be strengthened by precipitation heat treatment. The latter behaviour combined with the spherical particle shape confers the studied material higher strength and better machinability in comparison with similar composites produced using standard angular alumina particles. The overall features are promising for applications in the aerospace industry, where light and strong materials are required.

# 1. INTRODUCTION

Metal matrix composites (MMCs) have generated interest in academia and industry for more than four decades because of the possibility to tailor their physical and mechanical properties. According to [1] the global production of aluminium MMCs was 2700 metric tons in 2008. Figure 1 illustrates the dominance of ground transportation (mostly automotive) in terms of the produced volume of these materials. Due to high production rate requirements of the automotive industry, the preferred production methods in this segment are cost-effective squeeze-casting or pressure die-casting processes [2], which enable the material cost to be low enough for adoption by car manufacturers in real applications.

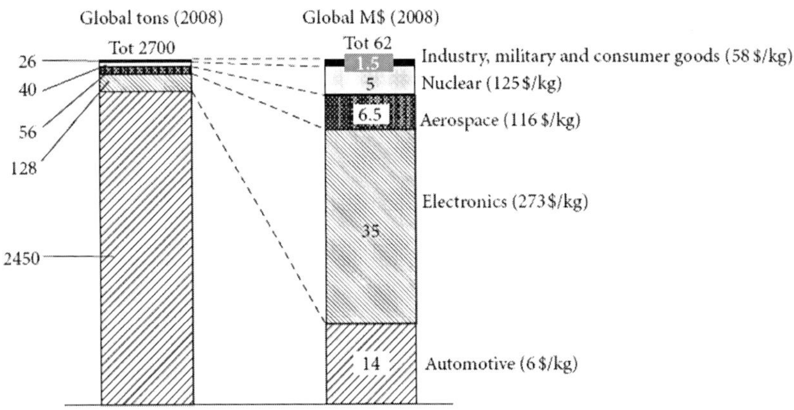


**Figure 1.** Global mass and value distribution amongst the main market segments for aluminium MMCs. Based on data from [1].

From a market value perspective, a remarkable 56% is captured by the electronics segment. The increasing importance of wireless communication and increasing power of microprocessors make electronics cooling of vital importance, typically to extend the lifetime and to allow integration in plastic housings (e.g., handset devices). The ability to tailor the coefficient of thermal expansion (CTE) and thermal conductivity has brought a real step change in designing electronics, driving the use of MMCs in this application. According to Mortensen [2], the preferred manufacturing process in this case is gas pressure infiltration because complexity and accuracy are required.

Aerospace applications represent 11% of the value. Composites for this industry are mainly processed by powder metallurgy, habitually followed by hot working, that is, forging, extrusion or rolling [1, 3]. In this case MMCs are used for parts that require high stiffness or high strength together with wear and fatigue resistance. In addition, other properties like toughness, corrosion resistance, or creep resistance (for warm and high-temperature applications) must meet required standards.

Aerospace manufacturers are constantly looking for lightweighting in order to reduce the fuel consumption. Every kg saved on a commercial aircraft corresponds to 700$ [1] saved for the airlines (whilst, for instance, the figure is 7$ for every kg saved, in the automotive sector); aluminium MMCs are therefore good candidates to further decrease these costs and turn aviation into a greener business.

The work presented in this paper addresses the mechanical and machining properties of an aluminium-based MMC reinforced with 60% volume fraction of spherical alumina particles. Its microstructure is shown in Figure 2. The composite of this work is obtained by gas pressure infiltration of an AA6061 alloy into a preform of packed spherical $Al_2O_3$ particles. The gas infiltration process allows producing highly reinforced, defect-free composites with the possibility to produce near-net shape components, and this potentially at lower production cost than using the powder metallurgy route. Angular alpha alumina, which is originally produced for abrasive products, has been and still is widely used for particle reinforced MMC (PRMMC) fabrication. This kind of powder is produced by comminution, which results in very sharp particles. A portion of these particles even contains cracks. When introduced into a metal matrix the result is a brittle composite [4]. In addition to that, during the composite processing (be it by squeeze casting, infiltration, or powder metallurgy), alpha alumina reacts with the magnesium contained in the present matrix alloy by forming a spinel $MgAl_2O_4$ phase at the matrix/particle interface. This reaction removes the Mg from the alloy [5, 6, 8]. This latter phenomenon reduces or cancels the strengthening effect obtained in the matrix through Mg-based precipitation. In addition, according to [6, 7], an embrittlement of the composite can occur due to the presence of this spinel phase. For all these reasons, the use of abrasive grade alpha $Al_2O_3$ powder as a reinforcement is not optimal.

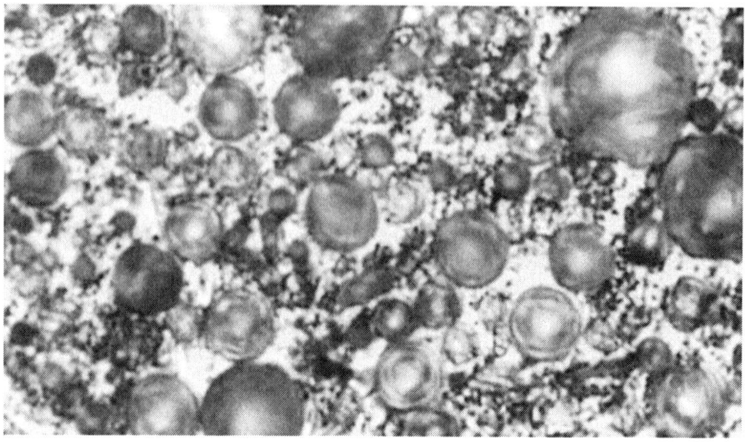

**Figure 2.** Microstructure of spherical PRMMC.

The choice of the reinforcement powder in the present paper has thus been driven by two main aspects: the nature of the alumina and the shape of the particles. The spherical ceramic powder selected here is produced and provided by GAP Engineering LLC based in Ash Grove, MO. As reported by Harrigan [8], this powder is an alpha alumina which has been modified into a mixed alpha-gamma-amorphous alumina. Its particularity is that during sintering it does not react with Mg additions to aluminium, thus retaining the strengthening effect in a matrix like AA6061 after precipitation heat treatment [9].

Composites of aluminium alloys reinforced with such spherical alumina particles are now produced commercially by powder metallurgy; they have attractive mechanical properties coupled with superior machinability compared with other particle reinforced aluminium composites [8, 9]. In work to date, the volume fraction of spherical particles has been kept in the range from 5 to 20%. Here, we explore significantly higher fractions ceramic, close to 60%. We do so by producing the composites by pressure infiltration, a process relying on flow of the matrix in a liquid state, which is better suited than powder metallurgy to the production of highly particle reinforced metal matrix composites [2]. This processing method will allow confirming whether or not the GAP alumina powder remains inert when facing, in the present case, an Mg-containing alloy in liquid form.

## 2. MATERIAL PROCESSING

Main characteristics of the GAP spherical alumina particles employed for the production of composites of this work are given in Table 1. In addition to GAP powder, Alodur angular alpha alumina powder, produced by Treibacher Schleifmittel in Laufenburg, Germany, was also infiltrated for comparison purposes. This powder, of grit size F600, has almost the same particle size as the GAP powder, in Table 1. Note also the difference in density between the two types of alumina, which gives a significant advantage to the lighter GAP powder, as discussed in Section 4.

**Table 1.** Main features of produced spherical and angular particle MMCs.

| Composite designation | Spherical PRMMC | Angular PRMMC |
|---|---|---|
| Matrix | AA6061 | AA6061 |
| Reinforcement type | $\alpha$-$\gamma$-Amorphous $Al_2O_3$ | $\alpha$-$Al_2O_3$ |
| Reinforcement shape | Spherical | Angular |
| Reinforcement density [g/cm$^3$] | 3.75* | 3.95** |
| Particle size [$\mu$m] | | |
| D3 | 31 | 19 |
| D50 | 6.5 | 9.3 |
| D94 | 2 | 3 |
| Volume fraction of reinforcement ($V_f$) | 0.6 | 0.6 |

*Source: GAP Engineering, measured by University of Missouri. **Source: Treibacher Schleifmittel product datasheet.

Both composites have been processed as described in [2] using a laboratory gas pressure infiltration device. The molten AA6061 alloy was pushed into the alumina preform with an argon gas pressure of 6 MPa and then directionally solidified thanks to a copper chill at his bottom. The resulting billets are 36 mm in diameter and 110 mm in length. After infiltration the billets were then heat-treated to T4 (solution heat treatment 530°C during 2 hours followed by water quench), T6 (T4 followed by precipitation heat treatment 160°C during 18 hours), and O (annealing treatment 415°C during 3 hours followed by cooling at 0.4°C/min till 260°C) tempers. Composites with sieved GAP powder have also been processed in order to observe eventual improvements on mechanical properties and machining. The sieving is obtained by letting the powder vibrate and fall through 20 µm and 10 µm grit sieves. Thus two powder batches could be produced, one with particles size <10 µm and another with particle size <20 µm.

# 3. TESTING

## 3.1. Mechanical Testing

Vickers hardness testing was carried out on a Brickers-200 machine from Gnehm (Horgen, Switzerland) by applying a 10 kg load during 16 s.

To measure elastic properties an impulse excitation apparatus was employed. The impulsion and the corresponding frequential response were, respectively, induced and captured by a GrindoSonic Mk5-Industrial device (J. W. Lemmens N. V., Leuven, Belgium) following the ASTM E1876 procedure.

Tensile tests were performed on rectangular section dogbone tensile specimens at a nominal strain rate of $3.10^{-5}\,s^{-1}$ according to ASTM E8 standard, using a MTS (Minneapolis, USA) Alliance RT/50 universal mechanical testing apparatus. Fractography and EDX mapping of the broken tensile samples were run with a scanning electron microscope (SEM, model FEI XL30-EBSP).

Tensile tests at 250°C were carried out at Liebherr Aerospace in Toulouse, France. In that case the strain rate was $8.10^{-5}\,s^{-1}$ in the elastic zone and $8.10^{-4}\,s^{-1}$ in the plastic zone in accordance with ASTM E21.

In order to compare the measured properties to relevant benchmarks, data for commercial Duralcan and Boralcan composites, matrix alloy AA6061, typical aerospace alloy AA2024, and AIRWARE 2196 (new lithium containing aerospace grade alloy) will be displayed on the relevant plots.

## 3.2. Machining

Preliminary qualitative turning trials were conducted with a machining expert from JRS Tools company (Payerne, Switzerland) in order to identify the right turning parameters and to find the appropriate insert material, geometry, and coating. This specific testing was set up to address one commonly mentioned barrier to the adoption for metal matrix composites, namely, the machining cost (the goal being to use general performance inserts and not expensive diamond cutting tools).

An $Al_2O_3$+TiC insert (grade HC4) from NTK Cutting Tools (Ratingen, Germany) with a TiN coating and a negative rake angle was selected. Machining trials were carried out on a W570E lathe from Voest-Alpine (Linz, Austria) with the following cutting parameters: cutting speed = 50 m/min, feed = 0.3 mm/rev, cutting depth = 0.5 and 2 mm (see next section), and dry machining. The chosen tool insert geometry as well as the cutting parameters is typical for rough turning, the goal being to reproduce high material removal machining rather than finishing.

Tool insert flank wear measurements and machined surface observation were done after removal of 50 $cm^3$ of material.

Insert flank wear (VB) measurements were done with a macrocamera capturing pictures of the worn zone and the machined surface was observed with a scanning electron microscope (SEM, model FEI XL30-EBSP).

# 4. RESULTS

## 4.1. Mechanical Properties: Hardness and Stiffness

Table 2 shows the measured values and compares the features of the spherical and angular particle reinforced composites. Elastic (E) and shear (G) modulus values are identical whilst the densities are slightly different due to a ~5% density gap between Alodur and GAP alumina particles. This difference becomes very important when calculating the specific bending stiffness, as illustrated in Figure 6. The hardness measurements reveal that precipitation of the hard phase $Mg_2Si$ in AA6061 matrix does not take place in the angular PRMMC, whereas a strengthening phenomenon brings the hardness from 167HV10 at T4 temper to 216HV10 at T6 temper in the spherical PRMMC. EDX mapping on the fractured surface and on the polished cross section, Figures 3(a) and 3(b), shows that magnesium is homogeneously distributed. No magnesium agglomeration around the particles is observed indicating that reaction between alumina and Mg does not happen, consistent with the heat treatment response, which suggests that Mg stayed in the matrix and was present to form precipitates. These observations confirm findings from [9] proving that the spherical alumina from GAP does not react with Mg. This proves true even in a liquid state process as the infiltration method used in the present work. Therefore, two major advantages of the novel powder can be identified for the spherical PRMMC: (i) Mg stays in the matrix and contributes to strengthen the composite by formation of precipitates, and (ii) formation of the brittle $MgAl_2O_4$ spinel at the particle-matrix interface is avoided.

**Table 2.** Main characteristics of spherical and angular composites.

| | Spherical PRMMC 60% | Angular PRMMC 60% |
|---|---|---|
| Density [g/cm$^3$] | 3.25 | 3.34 |
| $E$ [GPa] | 150 | 150 |
| $G$ [GPa] | 58 | 58 |
| Poisson's ratio | 0.29 | 0.29 |
| Hardness at T6 [HV10] | 216 | 159 |
| Hardness at T4 [HV10] | 167 | 165 |

The high reinforcement volume fraction of spherical PRMMC and angular PRMMC gives the composites an elastic modulus of 150 GPa. This is close to double the stiffness of commercial aluminium composites and alloys (Figure 4). The spherical PRMMC also shows the highest hardness. The combination of these properties makes it a potentially interesting material for structural applications in which low weight is a design target.

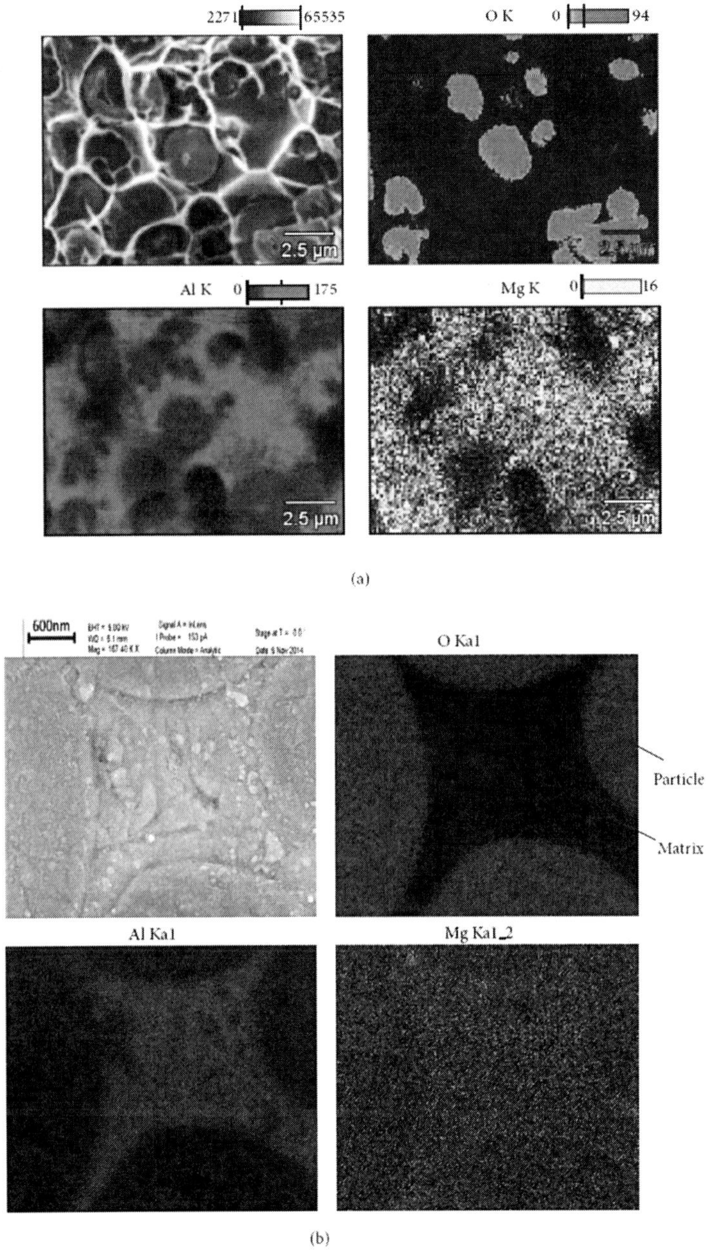


**Figure 3.** SEM imaging and EDX mapping for oxygen, aluminium, and magnesium of fracture surface (a) and polished surface (b) done on the spherical PRMMC.

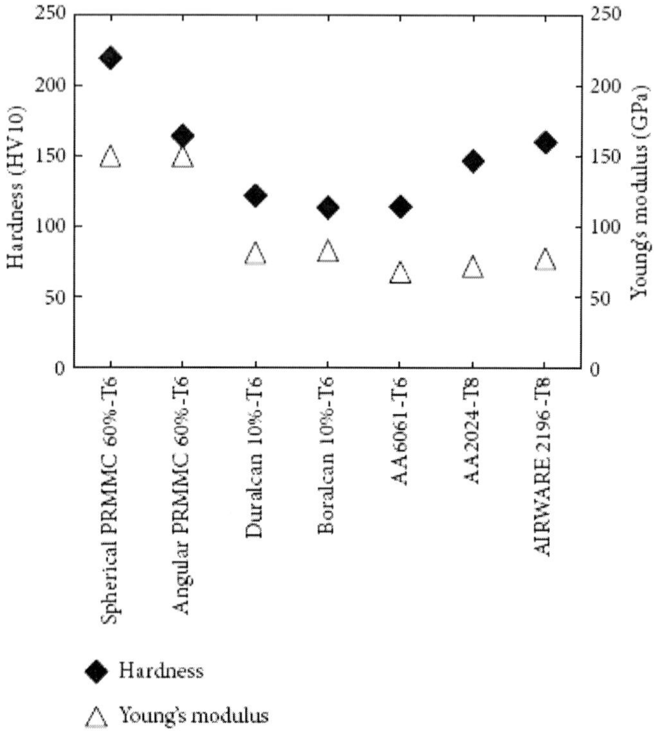


**Figure 4.** Hardness and elastic modulus comparison between spherical and angular PRMMC, commercial MMCs, reference AA6061 alloy and aerospace alloys AA2024, and AIRWARE 2196. All hardness values: measured. Young's modulus of spherical and angular PRMMC and Boralcan: measured. Young's modulus of Duralcan from [10], AA6061 and AA2024 from [11], and AIRWARE 2196 from Constellium product datasheet.

In aerospace applications low weight is of vital importance. Therefore, the density of the material has to be taken into account when evaluating its performance. Figure 5 illustrates that the highly reinforced composites have a high specific stiffness (/); however one also sees how the gap between them and the other materials is rapidly closed when calculating the rigidity in bending mode (/). The latter value is more meaningful as a figure of merit for a material in aerostructures where bending is the typical loading mode.

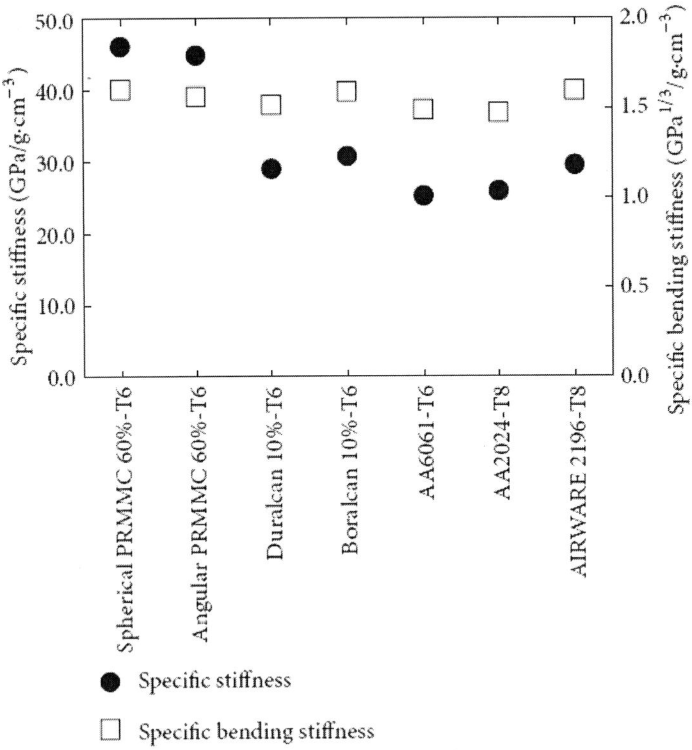


**Figure 5.** Specific stiffness and specific bending stiffness comparison between spherical and angular PRMMC, commercial MMCs, reference AA6061 alloy and aerospace alloys AA2024, and AIRWARE 2196. Densities for specific stiffness calculation: spherical and angular PRMMC from Table 2, Duralcan from [10], Boralcan from measurements, AA6061 and AA2024 from [12], and AIRWARE 2196 from Constellium product datasheet.

In order to better describe the differences in bending stiffness plotted in Figure 5 the relative improvement or reduction in bending stiffness versus the reference (unreinforced) alloy AA6061 is represented in Figure 6. The spherical PRMMC is 7.9% stiffer in bending than AA6061 alloy and also 2.9% better than the similarly processed angular particulate composite thanks to the slightly lower density of GAP alumina. That fact places the infiltrated GAP particle reinforced composite in the same range as best-in-class Al-lithium alloy AIRWARE 2196 and Boralcan composite, which contains light $B_4C$ particles ( = 2.52 g/cm$^3$).

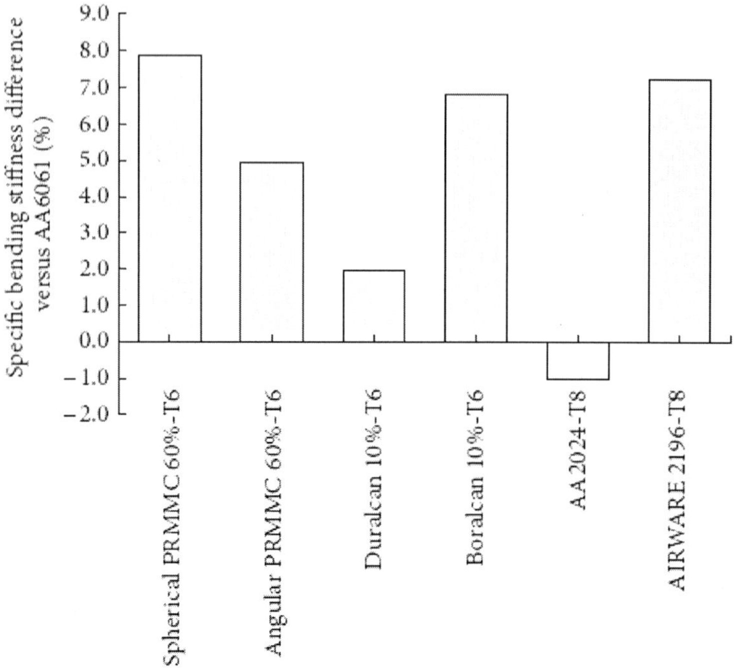


**Figure 6.** Bending stiffness difference relative to unreinforced AA6061 reference alloy.

## 4.2. Mechanical Properties: Tensile

Tensile tests have been performed on the spherical alumina particle reinforced MMC, comparing the mechanical properties for different metallurgical states of the matrix: maximizing strength with the T6 temper or maximizing ductility with the O temper. The curves obtained, in Figure 7(a), show a good repeatability, except for variability regarding the elongation at fracture; this is in turn linked to the rather brittle fracture mode of the composite. This brittle behaviour is confirmed by low elongation values in the T6 condition (Table3) which makes the application of such a material problematic in aerospace. A higher elongation value of 1.35% is obtained when the spherical PRMMC is in O condition. In that case the yield strength (YS) and ultimate tensile strength (UTS) are 238 MPa and 330 MPa, respectively. A rule of thumb for aerospace materials is to have YS above 300 MPa with an elongation above 1%. The present composite does not meet this requirement yet is sufficiently close to warrant an exploration of its thermomechanical postprocessing in future work.

**Table 3.** Mechanical properties of spherical PRMMC at room temperature (non aged) and at 250°C after 1000 h ageing at 250°C.

| Test temperature [°C] | Temper | YS [MPa] | | UTS [MPa] | | El. [%] | | E [GPa] |
|---|---|---|---|---|---|---|---|---|
| | | Average | Maximum | Average | Maximum | Average | Maximum | |
| 25 | T6 | 413 | 415 | 447 | 457 | 0.44 | 0.55 | 150 ± 8 |
| 25 | O | 238 | 239 | 330 | 337 | 1.25 | 1.35 | 150 ± 8 |
| 250 | T6 | 195 | 201 | 210 | 217 | 2.50 | 2.70 | 105 ± 7 |

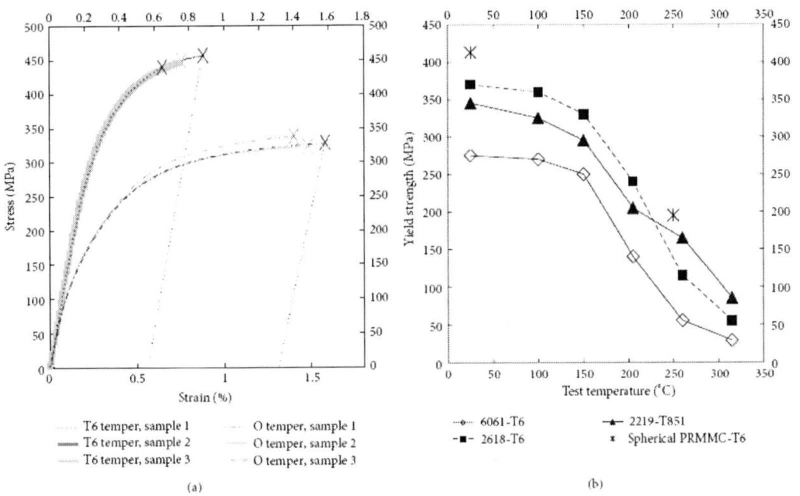


**Figure 7.** (a): Tensile curves of spherical PRMMC in T6 and O tempers at 25°C. (b) Yield strength evolution between 25°C and 315°C of AA6061 and high-temperature alloys AA2618 and AA2219 [11] as well as spherical PRMMC values at 25°C and 250°C. All >50°C data are after 1000 h ageing at the test temperature.

The use of smaller size GAP powder particles to enhance elongation has been attempted by producing composites with sieved reinforcement (all particles <10 µm or <20 µm in size). These variants did not show significant difference in tensile properties, sieving to such sizes thus being ineffective for static properties. Control of particles size is however expected to be more beneficial for fatigue resistance by postponing the crack initiation and decreasing the frequency of particle fracture during crack propagation [13, 14]. Additionally, a smaller particle size in high reinforced composites reduces strain-rate sensitivity and damage accumulation during dynamic compression and therefore is potentially beneficial for postprocessing like forging [15, 16].

Further measurements were performed in collaboration with Liebherr Aerospace in Toulouse, France. The spherical PRMMC was exposed to 250°C during 1000 h with subsequent tensile testing at 250°C. In those conditions YS and UTS drop to 195 and 210 MPa, respectively, and the elongation reaches a value of 2.7%. The influence of high reinforcement volume fraction is significant and beneficial when comparing yield strength of AA6061 matrix alloy with spherical PRMMC at 250°C. This is illustrated in Figure 7(b), which also includes curves of current high-temperature aluminium alloys AA2618 and AA2219. The spherical PRMMC shows the highest yield strength value at 250°C. This confirms its potential for warm/hot applications.

These values can be interpreted as a further step in "softening" the matrix to obtain higher elongations and show that the value of 2.7% can be physically reached for such highly reinforced material provided its matrix has adequate properties.

SEM observation of the fracture surfaces reveals that the tested composite is free of broken or cracked particles (Figure 8). Most particles stay embedded in the matrix after fracture. This suggests the presence of a rather healthy interface (no porosity, no microcracks, and no spinel phase). Final material separation happens to a large extent through the thin interparticle matrix bridges.

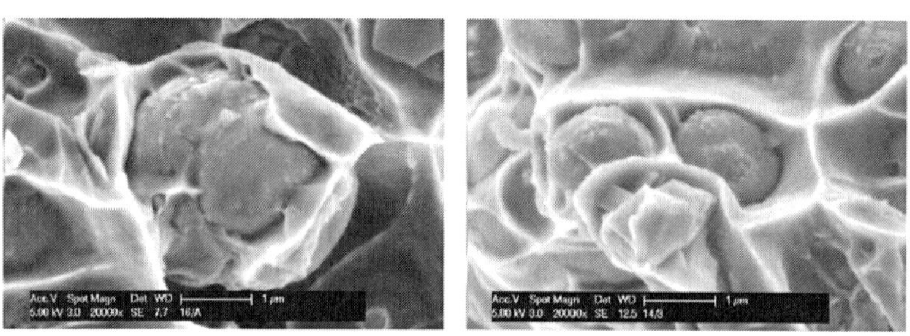

**Figure 8.** Fracture surfaces of the spherical PRMMC showing the tight bonding between the particles and the matrix. Magnification 20000x.

Strong particles and defect-free interfaces combined with the spherical shape thus allow such high reinforced composite to show an interesting balance of strength and ductility. Improvements of the matrix by applying adequate treatments or tuning its composition can strengthen the network of interparticle bridges and allow the properties of the MMC to match the requirements for aerospace applications.

## 4.3. Machining

To be used in an actual application, the aptitude of the material to be shaped is an important factor. Powder metallurgy composites reinforced with GAP powder at lower loadings than in the present work have been tested for machining properties by Harrigan [9], showing significant improvements in tool life compared to equivalent SiC composites. This observation was made on 15% reinforced composites, using TiN coated tungsten carbide inserts.

The same procedure was employed in this work in order to evaluate the wear of the tool insert when machining the 60 vol% GAP powder reinforced composite. Figure 9 shows that the highly reinforced spherical PRMMC wears dramatically the tool insert, mainly by two modes: flank wear and cratering. Flank wear cannot by definition be avoided, while cratering typically appears when machining abrasive materials.

The first round of trials (Figure 9(b)) using a cutting depth of 0.5 mm resulted in a flank wear (VB) of 1.15 mm. This is almost four times higher than the threshold value of 0.3 mm. Above this threshold value the tool insert is considered to be no longer usable [17]. By increasing the cutting depth to 2 mm, the insert flank wear value decreased to reach 0.8 mm. This result, obtained through a selection of machining conditions in addition to the spherical shape of the particles [9], brings the spherical PRMMC into the wear range of standard and low reinforced composites as illustrated in Figure 10. This is an encouraging result, although more trials and parameter exploration have to be done in order to come closer to VB = 0.3 mm.

Concerning the behavior of the composite during machining, the SEM image in Figure 12(a) shows that particles do not break and are not torn out whilst the tool insert removes the material. Cracks on the machined surface are not observed either. These observations confirm the positive effect of spherical particle shape on machining as explained by Harrigan

[9]. The feed motion and the negative rake angle (Figure 11) produce compressive forces on the MMC. Under compression the spherical particles tend to be pushed and embedded into the matrix. The combination of the latter with the spherical particle effect allows such a high reinforced MMC to be machined without major surface damage and with limited tool insert wear.

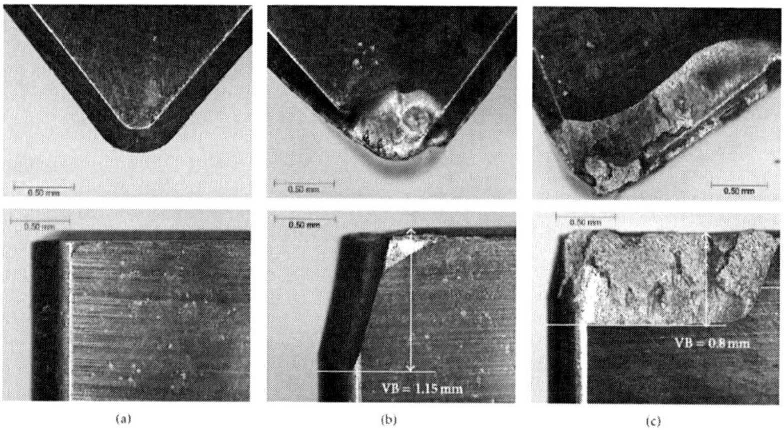


**Figure 9.** (a) New tool insert. (b) Worn tool insert after removing 50 $cm^3$ of material by turning, cutting depth = 0.5 mm. (c) Worn tool insert after removing 50 $cm^3$ of material by turning, cutting depth = 2 mm.

Figure 12(b) shows the trace left by the tool insert on the machined surface. Differences in roughness can clearly be distinguished between region M and region N. Compression due to combination of feed motion and negative rake angle gives a smooth and void free surface in region M. The absence of compression from feed motion behind the cutting nose generates region N where voids are formed and some particles get less embedded than in region M.

This confirms the importance and the need of finishing machining to avoid the presence of these cavities, which represent potential crack initiation sites.

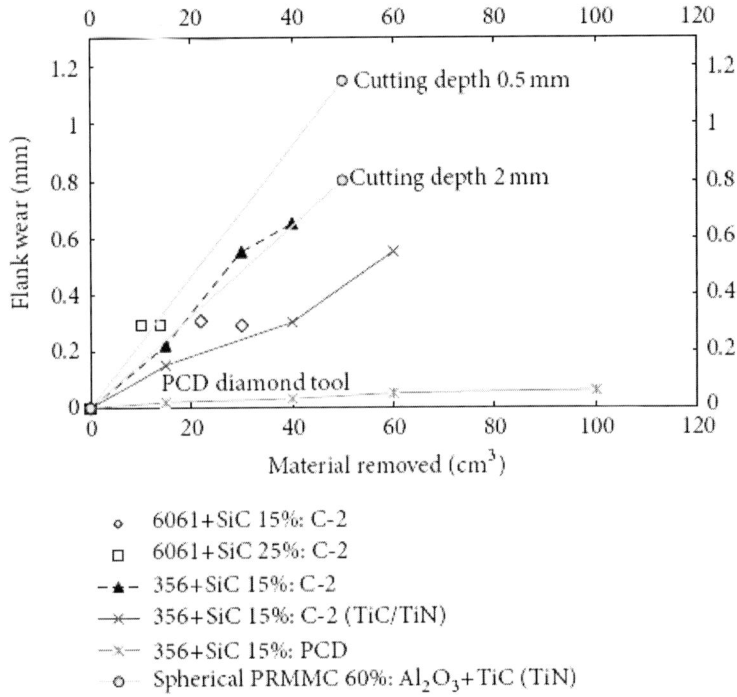


**Figure 10.** $Al_2O_3$+TiC (coated with TiN) insert flank wear after machining spherical PRMMC, comparison with five other MMCs [9], machined with noncoated carbide (C-2), TiC/TiN coated carbide, and PCD diamond inserts.

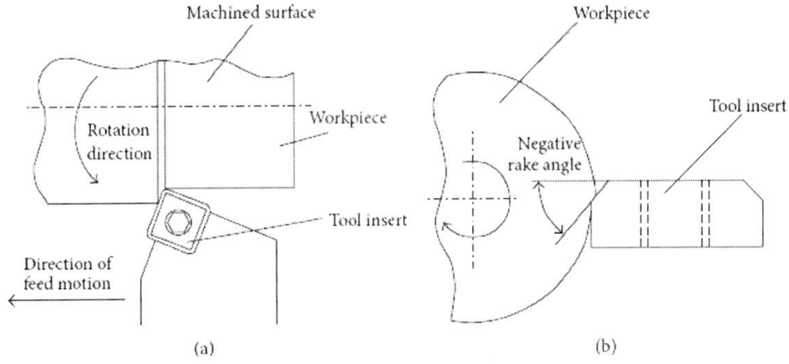


**Figure 11.** Schematic illustration of turning. (a) Top view. (b) Front view.

**Figure 12.** (a) SEM capture of machined surface, magnification 2000x. (b) SEM capture of machined surface divided in two regions. M: smooth and void-free surface and N: rough surface with cavities. Magnification 250x.

# 5. CONCLUSIONS

1) Dissimilar behaviours are observed between standard alpha alumina and alpha-gamma-amorphous GAP alumina when used as reinforcement in AA6061 matrix alloy. Hardness tests at T4 and T6 tempers reveal that precipitation hardening is possible in GAP alumina reinforced 6xxx series aluminium alloys. This is because no reaction between magnesium and alumina happens during infiltration, solidification, or cooling of the composite. This behaviour is the opposite of what is observed with alpha alumina particles.

2) Tensile tests show that the ductility and strength tradeoff is only slightly below the critical threshold for applicability of the composites. In aerospace practice a minimum of 300–350 MPa for YS and 1% for elongation is required as a base line. In order to meet these targets an optimum temper/treatment or an alternative matrix has to be found.

3) Wear of tool inserts is very high when turning such highly reinforced composite. Nevertheless, adapted machining parameters combined with the spherical shape of the alumina particles allow being in the same range of wear of low volume fraction SiC particle reinforced aluminium composites.

4) Gas pressure infiltration processing offers the possibility to obtain pore-free near-net shape pieces with high elastic modulus. The combination of the latter with further improvements in composite design, postprocessing, and machining procedures may allow the application of highly reinforced MMCs and let aerospace applications benefit from their high specific properties.

## ACKNOWLEDGMENTS

The authors would like to thank contributors from the Ecole Polytechnique Fédérale de Lausanne, that is, the Laboratory of Mechanical Metallurgy and the Mechanical Workshop of Material Science Institute ATMX for enabling the production and testing of the MMCs as well as Professor Andreas Mortensen for his expertise in the matter. Furthermore, Liebherr Aerospace is thanked for the high-temperature tensile tests and Dr. William Harrigan for the information on GAP powder and on machining of MMCs.

## REFERENCES

1. C. Swift, "Metal matrix composites: the global market," Market Research Report, BBC Research, 2009.

2. A. Mortensen, "Melt infiltration of metal matrix composites," in Comprehensive Composite Materials, Volume 3: Metal Matrix Composites, T. W. Clyne, Ed., chapter 3.20, pp. 521–554, Pergamon Press, New York, NY, USA, 2000.

3. D. B. Miracle, "Metal matrix composites—from science to technological significance," Composites Science and Technology, vol. 65, no. 15-16, pp. 2526–2540, 2005.

4. A. Miserez and A. Mortensen, "Fracture of aluminium reinforced with densely packed ceramic particles: influence of matrix hardening," Acta Materialia, vol. 52, no. 18, pp. 5331–5345, 2004.

5. A. D. McLeod and C. M. Gabryel, "Kinetics of the growth of spinel, $MgAl_2O_4$, on alumina particulate in aluminum alloys containing

magnesium," Metallurgical Transactions A, vol. 23, no. 4, pp. 1279–1283, 1992.

6. S. Vaucher and O. Beffort, Bonding and Interface formation in Metal Matrix Composites, vol. 9, MMC-Assess Thematic Network, EMPA, 2001.

7. J. C. Lee, K. N. Subramanian, and Y. Kim, "The interface in $Al_2O_3$ particulate-reinforced aluminium alloy composite and its role on the tensile properties," Journal of Materials Science, vol. 29, no. 8, pp. 1983–1990, 1994.

8. W. C. Harrigan, "Machinable aluminium matrix composites," in Proceedings of the Aeromat 2011 Presentation, 2011.

9. W. C. Harrigan, "Machinable aluminium matrix composite," in Conference Proceedings: Commonality of Phenomena in Composite Materials II, AIME, 2011.

10. A. Evans, C. S. Marchi, and A. Mortensen, Metal Matrix Composites in Industry: An Introduction and a Survey, vol. 1, Kluwer Academic Publishers, 2003.

11. J. G. Kaufman, Properties of Aluminium Alloys: Tensile, Creep, and Fatigue Data at High and Low Temperatures, AA & ASM International, 1999.

12. Pechiney-Rhenalu, Aluminium Semi-Finished Products, Pechiney-Rhenalu, 1997.

13. J. N. Hall, J. Wayne Jones, and A. K. Sachdev, "Particle size, volume fraction and matrix strength effects on fatigue behavior and particle fracture in 2124 aluminum-$SiC_p$ composites," Materials Science and Engineering A, vol. 183, no. 1-2, pp. 69–80, 1994.

14. D. A. Lukasak and D. A. Koss, "Microstructural influences on fatigue crack initiation in a model particulate-reinforced aluminium alloy MMC," Composites, vol. 24, no. 3, pp. 262–269, 1993.

15. A. Awadallah and J. J. Lewandowski, "Forging of discontinuously reinforced aluminum composites," inASM Handbook, Volume 14A—Metalworking: Bulk Forming, pp. 366–373, 2005.

16. C. San Marchi, F. Cao, M. Kouzeli, and A. Mortensen, "Quasistatic and dynamic compression of aluminum-oxide particle reinforced pure aluminum," Materials Science and Engineering A, vol. 337, no. 1-2, pp. 202–211, 2002.

17. V. P. Astakhov and J. P. Davim, "Tools (geometry and material) and tool wear," in Machining: Fundamentals and Recent Advances, pp. 29–57, 2008.

8

# MICROGALVANIC CORROSION BEHAVIOR OF CU-AG ACTIVE BRAZE ALLOYS INVESTIGATED WITH SKPFM

**Armen Kvryan, Kari Livingston, Corey M. Efaw, Kyle Knori, Brian J. Jaques, Paul H. Davis, Darryl P. Butt and Michael F. Hurley ***

Department of Materials Science and Engineering, College of Engineering, Boise State University, Boise, ID 83725-2090, USA

## ABSTRACT

The nature of microgalvanic couple driven corrosion of brazed joints was investigated. 316L stainless steel samples were joined using Cu-Ag-Ti and Cu-Ag-In-Ti braze alloys. Phase and elemental composition across each braze and parent metal interface was characterized and scanning Kelvin probe force microscopy (SKPFM) was used to map the Volta potential differences. Co-localization of SKPFM with Energy Dispersive Spectroscopy (EDS) measurements enabled spatially resolved correlation of potential differences with composition and subsequent galvanic corrosion behavior. Following exposure to the aggressive solution, corrosion damage morphology was characterized to determine the mode of attack and likely initiation areas. When exposed to 0.6 M NaCl, corrosion occurred at the braze-316L interface preceded by preferential dissolution of the Cu-rich phase within the braze alloy. Braze corrosion was driven by galvanic couples between the braze alloys and stainless steel as well as between different phases within the braze microstructure. Microgalvanic corrosion between phases of the braze alloys was investigated via SKPFM to determine how corrosion of the brazed joints developed.


## KEYWORDS

active braze alloy; corrosion; microgalvanic corrosion; electrochemistry; scanning kelvin probe force microscopy; stainless steel; Cu-Ag alloy; joining; scanning electron microscopy; energy dispersive spectroscopy


## 1. INTRODUCTION

Failure, whether chemical or mechanical in nature, often occurs where dissimilar materials are joined together. Common joining techniques range from mechanical fastening (e.g., bolt or rivet) to solid state joining (e.g., diffusion bonding). Brazing is an alternative joining technique in which a filler material is heated so as to form metallurgical bonds with the surfaces of the parts (parent materials) being joined. Brazing is a relatively low temperature process wherein the filler (braze material) is heated to a temperature above its melting point but below that of the parent materials. Joints formed by brazing can achieve very tight tolerances and offer desirable mechanical properties, similar to diffusion bonded materials, but have the advantage of being easily disassembled [1].

Many different braze alloys are available—designed for specific applications. A braze that performs well mechanically may have limited utility based on environmental compatibility. For example, silver-based braze alloys typically have a lower corrosion rate in an industrial atmosphere than in a marine environment [2]. When used in settings where contact with certain types of fuel may occur, silver-based brazes have poor corrosion resistance that can ultimately lead to failure [3]. The effect of brazing on the corrosion behavior of the resultant parts (including parent materials) has not been widely investigated. Through gaining a better understanding of this vital relationship between brazing and its influence on corrosion behavior, it may be possible to optimize processing parameters and develop more effective brazes. Thus, it is important to understand the mechanism(s) that control the corrosion development at and near joints. Consequently, the impact of brazing on the corrosion behavior of the overall system is investigated in this work.

## 1.1. Background

The Cu-Ag based systems investigated in this study were chosen because of their ease of application, commercial availability, and favorable compatibility with common engineering metals and technical ceramics. Good materials compatibility is due in part to effective wetting. Wetting, the ability of a liquid to maintain contact with a solid surface, is an important factor in the overall effectiveness of a braze [4]. The wetting parameter, or wettability, is the geometric complement of the contact angle, and thus is inversely related to the contact angle [5]. During brazing, good compatibility is achieved when the braze alloy wets the target material well. This ensures excellent surface coverage and any diffusion that occurs does not result in poor mechanical behavior. To promote wetting and joint compliance, reactive elements are added to brazing alloys [1]. The resultant brazing alloys are known as active brazing alloys (ABA). Titanium is often added to silver based braze alloys in order to enhance their ability to wet the parent material(s) being joined, which is vital when joining dissimilar materials, including ceramics [1]. In addition, an increased amount of titanium in brazes has been found to yield thicker joints [6]. In this study, commercially available Cu-Ag-Ti and Cu-Ag-In-Ti braze alloys were obtained and used to join 316L stainless steel samples. It should be noted that oxide layers on stainless steel, including 316L, can act as a barrier to wetting, resulting in lower degrees of wettability for the braze [7]. The indium additions (~12%) present in the Cu-Ag-In-Ti braze lower the metal surface tension [8]. Additionally, indium additions in a titanium-containing braze alloy increase activity of titanium and wettability [8,9].

In any joining technique, the joint must preserve mechanical integrity and environmental compatibility. Heat input and the introduction of dissimilar materials may provide initiation spots for pitting corrosion at microstructural heterogeneities, galvanic attack at the metal/braze interface, or dealloying within the braze material. Initially, corrosion within the braze alloy may cause loss of joint integrity and sustained braze corrosion may create an aggressive local chemical environment that can lead to depassivation of the parent material. Relatively low temperature brazing alloys typically rely on noble metals to provide good wetting behavior, adhesion, and joint compatibility. However, these alloys generally provide a thermodynamic driving force for galvanic corrosion when joined with the parent metal and are exposed to conditions that support active corrosion. Galvanic corrosion as the mechanism of failure for silver based brazes coupled to stainless steels was first proposed by Takemoto *et al.* [10].

Current research on the effect of brazing has mainly focused on compatibility and resultant mechanical behavior [11,12,13,14]. However, once a candidate braze has been identified, the reliability and resistance to environmental attack must be addressed to determine long-term viability. With the exception of a few isolated studies [15,16,17,18], the effect of brazing on corrosion behavior has received little attention in the open literature [18]. Because of the multicomponent aspect of brazes, a multiphase microstructure is usually seen post brazing where the two parent materials are joined [18]. This is mainly due to precipitation of stable or meta-stable phases during the brazing cycle [16]. Once these phases are present, they would be expected to have different electrochemical behavior based on composition. The local potential difference between dissimilar regions in the microstructure will influence corrosion behavior, and is known as microgalvanic corrosion. The term microgalvanic corrosion describes a galvanic corrosion cell occurring on a sub-grain scale. Regions of varying composition result in potential differences amongst the individual phases in the braze, and tend to increase the corrosion rate of the less noble phases [16]. The concept of microgalvanic corrosion is important to joint reliability, as manipulations of the filler metal in the braze can dramatically influence the overall corrosion behavior of a braze [19].

Confirmation of a microgalvanic corrosion mechanism requires accurate characterization of the surface of the materials, both in terms of composition and electrochemical potential. Scanning Kelvin Probe Force Microscopy (SKPFM) has been shown to be an effective technique to characterize expected electrochemical behavior of surface inhomogeneities within a metal alloy's microstructure [20]. A Kelvin probe measures the work function difference (Volta potential) between the surface of a sample and the probe itself. Correlation between the Volta potential difference (VPD) obtained via SKPFM and electrode solution potentials, and hence the likely development of galvanic couples during active corrosion conditions has previously been established [21]. However, this relationship should be approached with some caution for unconfirmed systems. SKPFM is a surface technique and sensitive to the formation of surface reaction products such as oxide layers that may influence the measurement, and thus may not always directly correlate with solution potential. However, by using an Atomic Force Microscope (AFM) with surface potential feedback and nullification, SKPFM is able to map Volta potential differences on a surface with extremely fine (sub-micron) resolution. Compared to other local techniques, SKPFM currently provides the highest achievable spatial resolution for studying corrosion initiation driven by microstructure inhomogeneities. This paper utilized SKPFM to analyze and characterize the VPD between the different metallic/intermetallic phases present within the brazes studied. This method was used to investigate and explain corrosion initiation and propagation driven by galvanic corrosion arising from compositional differences within brazing alloys used to join 316L stainless steel.

# 2. MATERIALS AND METHODS

## 2.1. Materials and Joining Procedure

Commercial 316L stainless steel sheet with a thickness of 3 mm was water-jet cut into disk shaped samples of various sizes to be brazed. The braze alloys used were Cu-Ag-Ti-ABA with a thickness of approximately 50 μm and Cu-Ag-In-Ti-ABA with a thickness of 55 μm. Compositions of all materials are listed in Table 1 and Table 2. Stainless steel disks were polished to a 1 μm finish with SiC polishing pads. The as-received Cu-Ag-Ti and Cu-Ag-In-Ti foils were lightly polished to an 800 grit finish and cut to sizes just smaller than the stainless steel disks. After polishing, all samples were ultrasonically cleaned in ethanol for 15 min, washed in deionized water, and dried using compressed air.

**Table 1.** Compositions in atomic percent of the braze alloys.

| Material | Ag | Cu | Ti | In |
|---|---|---|---|---|
| Cu-Ag-Ti | 63.1 | 35.1 | 1.8 | – |
| Cu-Ag-In-Ti | 59.00 | 27.25 | 1.25 | 12.5 |

Two geometrical configurations were used during brazing: sandwich (stainless steel/braze foil/stainless steel) and coated (braze foil on stainless steel disk). Prior to firing, the cleaned and polished samples were placed in an alumina boat wrapped in niobium foil and inserted into the furnace hot zone. The system was then purged with ultra-high purity argon gas (UHP Ar, 99.999%, Norco, Boise, ID, USA) for 20 min before continuously flowing with oxygen gettered UHP Ar. Stainless steel joining with Cu-Ag-Ti was achieved through the following thermal cycle suggested by the supplier: ramp to 700 °C (80 °C below the solidus temperature of Cu-Ag-Ti) at 5 °C/min and hold for 20 min before ramping to 830 °C (15 °C above the liquidus temperature of Cu-Ag-Ti braze alloy) at a rate of 5 °C/min and held there for 15 min. The furnace was then cooled to room temperature at a rate of 5 °C/min. To create Cu-Ag-In-Ti joints, the furnace was ramped from room temperature to a temperature of 500 °C (130 °C below the solidus temperature of Cu-Ag-In-Ti) at a rate of 5 °C/min and held for 20 min, followed by a ramp to 730 °C (15 °C above the liquidus temperature of Cu-Ag-In-Ti) at a rate of 5 °C/min and held for 15 min, then cooled at a rate of 5 °C/min to room temperature, again as per the supplier recommendation.

**Table 2.** Compositions in atomic percent of the material being joined.

| Material | C | P | Ni | Cr | Mn | Mo | N | Si | S | Fe |
|---|---|---|---|---|---|---|---|---|---|---|
| 316L | 0.0002 | 0.00036 | 0.101 | 0.1663 | 0.0168 | 0.0203 | 0.00066 | 0.00485 | 0.00026 | Balance |

## 2.2. Scanning Kelvin Probe Force Microscopy (SKPFM) of Brazed Regions

For SKPFM, brazed samples were cross sectioned, cold mounted in epoxy, and polished with progressively finer grit silicon carbide pads and diamond slurries starting with 400 grit and ending with a 1 µm diamond slurry. Next, the samples were cleaned ultrasonically in a bath of non-denatured (200 proof) High Performance Liquid Chromatography (HPLC)/ spectrophotometric grade ethanol (Sigma-Aldrich, St. Louis, MO, USA). The final polishing step employed a VibroMet 2 vibratory polisher (Buehler, Lake Bluff, IL, USA) in which samples were polished for 1 h using a 12" Mastertex PSA pad (Buehler) covered with a 0.50 µm diamond slurry (MasterPrep Polishing Solution, Buehler). The polished samples were then rinsed with ethanol and blown dry with compressed air. Prior to imaging, an electrical connection between the sample surface and the AFM stage was established using colloidal silver paint and verified with a voltmeter.

Imaging was conducted using a Bruker Dimension Icon AFM operating in frequency modulation Peak Force KPFM (FM PF-KPFM) mode with a PFQNE-AL probe (Bruker, Santa Barbara, CA, USA). PF-KPFM is a dual-pass method wherein the first pass acquires topography via Peak Force tapping (*i.e.*, rapid force curves). The second pass is then used to measure the tip-sample surface potential difference at a user-determined fixed lift height above the sample surface. The SKPFM technique and important experimental considerations have been described in greater detail previously [22], but lift heights of ~100 nm and a frequency modulation based detection scheme were employed. To enhance signal to noise and minimize the effects of residual sample roughness on the surface potential image, a slow scan rate was used (~0.05–0.1 Hz). Because SKPFM measures the difference in work function between the AFM probe and the sample surface (*i.e.*, the relative rather than absolute surface potential), potentials of the different phases are reported relative to each other. Absolute potentials can be determined; e.g., using an inert gold standard as a reference material.

## 2.3. Braze Phase Composition

A Scanning Electron Microscope (SEM) (Hitachi S-3400N-II, Hitachi, Tokyo, Japan) equipped with energy dispersive spectroscopy (EDS) capabilities was used to image and analyze the mounted braze joint cross-sections for both Cu-Ag-Ti and Cu-Ag-In-Ti alloys after SKPFM characterization. Performing SEM after SKPFM was to avoid effects of electron beam irradiation on the surface potential measured. The cross section of each braze alloy was imaged in both secondary electron mode and backscatter electron mode. Subsequently, elemental mapping was performed on the cross sections of each sample, along with multi-point analysis at selected locations.

## 2.4. Corrosion Testing

To assess corrosion behavior electrochemical and exposure testing in 0.6 M NaCl solution at room temperature (22–24 °C) was conducted. Potentiodynamic testing was performed on both the braze alloys and stainless steel samples separately following thermal treatment using the prescribed brazing furnace profile. The braze alloy samples were prepared by applying multiple coatings of either the foil or paste braze compound to one face of a stainless steel coupon. Multiple braze coatings ensured that only the braze material was exposed to solution when tested, with no impact from the underlying stainless steel possibly present at the pinholes that may occur with a single coating. A conventional 3-electrode cell with a Pt mesh counter electrode and a saturated calomel (SCE) reference electrode was used for potentiostat controlled polarization measurements.

# 3. RESULTS

## 3.1. Brazed Stainless Steel Joint Characterization

Metallographic samples of the post-braze joint cross-section were examined via SEM in order to characterize the resulting joint (Figure 1). The joints formed a clean, tight, hermetic seal with the parent 316L samples. Both brazes were approximately 50 μm in width and displayed distinctive eutectic type phases in the braze region following the prescribed thermal cycles.

(a) (b)

**Figure 1.** Scanning Electron Microscopy (SEM) images of cross sections of the resultant braze joint between two stainless steel 316L samples joined using: Cu-Ag-Ti (**a**) and Cu-Ag-In-Ti (**b**) braze alloys.

EDS line scans are presented in Figure 2 (Cu-Ag-Ti) and Figure 3 (Cu-Ag-In-Ti) below. The Cu-Ag-Ti braze had a Ag-rich matrix with numerous distributed regions, generally ~1–10 μm across, of a precipitated Cu-rich phase. Generally, the EDS results did not indicate noticeable diffusion of the braze filler materials into the stainless steel. However, in a few areas thc EDS scans indicated some transport of the braze alloy elements into the stainless steel, resulting in a Ti and Ag-containing region adjacent to the braze for the Cu-Ag-Ti braze sample. It was also observed that that the bulk of the titanium in the braze segregated to one side of the braze/stainless steel interface (Figure 2). The Cu-Ag-In-Ti braze phase structure consisted of an Ag-In-rich matrix phase with numerous regions of a precipitated Cu-rich phase. Ti was also found to be present in some of the Cu-rich regions.

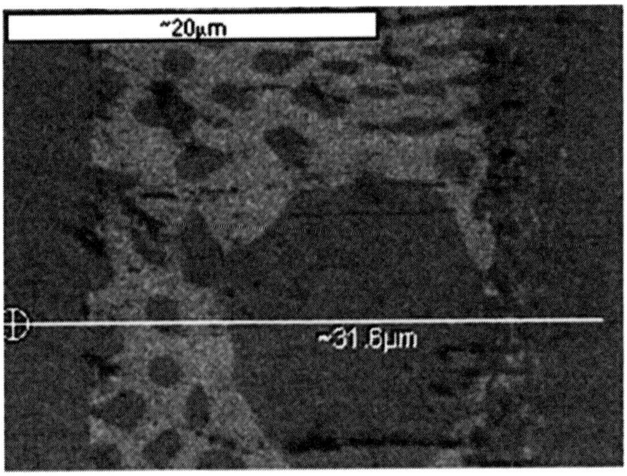


(a)

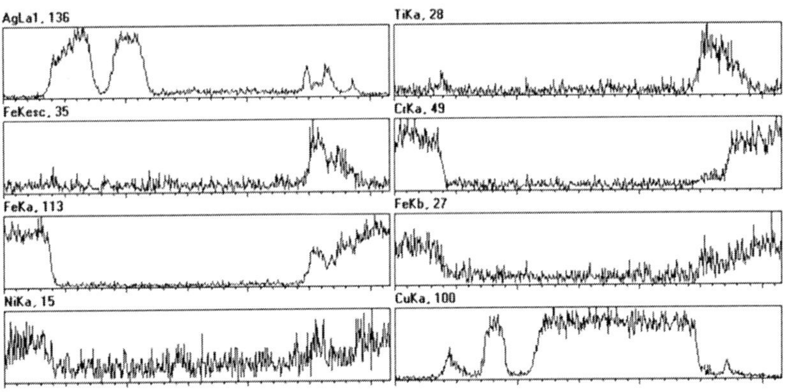


(b)

**Figure 2.** (**a**) SEM micrograph and (**b**) energy dispersive spectroscopy (EDS) line scan results from the cross section of a 316L stainless steel/Cu-Ag-Ti brazed joint.

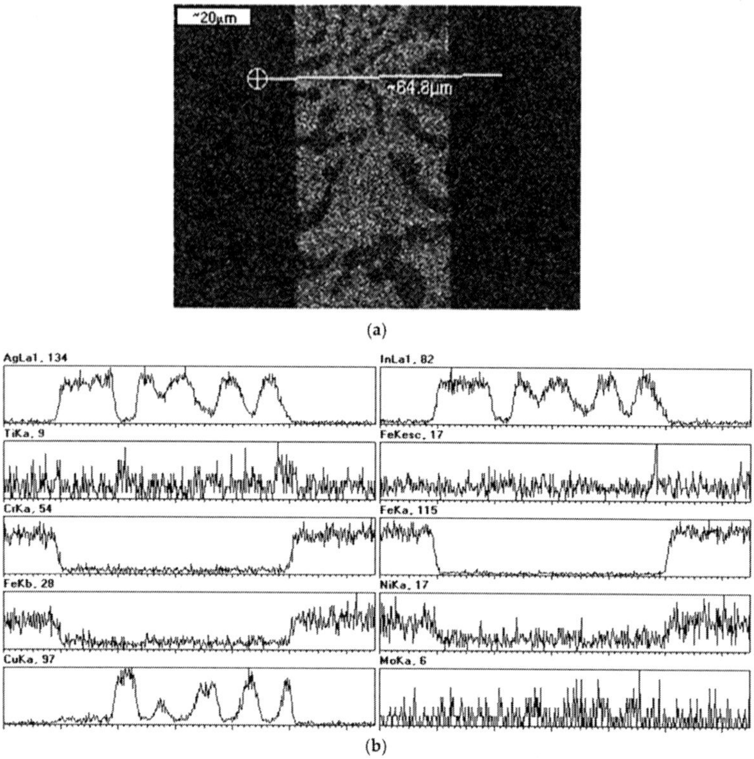


**Figure 3.** (**a**) SEM micrograph and (**b**) EDS line scans of a cross section of a 316L stainless steel/Cu-Ag-In-Ti brazed joint.

Phase separation within the braze regions was further characterized via optical microscopy to aid with identification of the various phases present when initiating SKPFM scans. Optical images obtained of the Cu-Ag-Ti and Cu-Ag-In-Ti braze areas showed clear variation of phase composition within the braze, with uniform 316L stainless steel on either side (Figure 4). The identical sample regions seen in Figure 4 were also characterized and discussed later in this paper.

In order to obtain semi-quantitative compositions of the phases, EDS multi point scans were acquired for both Cu-Ag-Ti and Cu-Ag-In-Ti samples. Multiple spectra were taken at representative points within the individual phase regions of each sample. The average composition values in atomic percent obtained for each phase present are listed in the tables adjacent to the respective images (see Figure 4). The Cu-Ag-Ti sample had a standard deviation below 3, whilst the Cu-Ag-In-Ti had a standard deviation below 5.

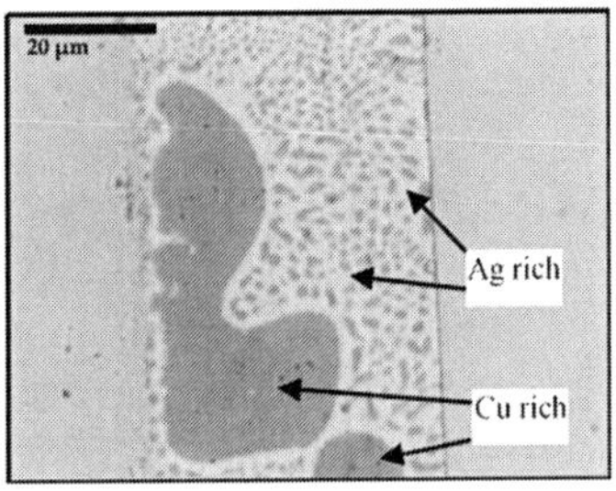


| Phase | Cu | Ag | Ti |
|---|---|---|---|
| Cu Rich | 94.65 | 5.10 | 0.50 |
| Ag Rich | 10.01 | 89.98 | 0.01 |

(a)

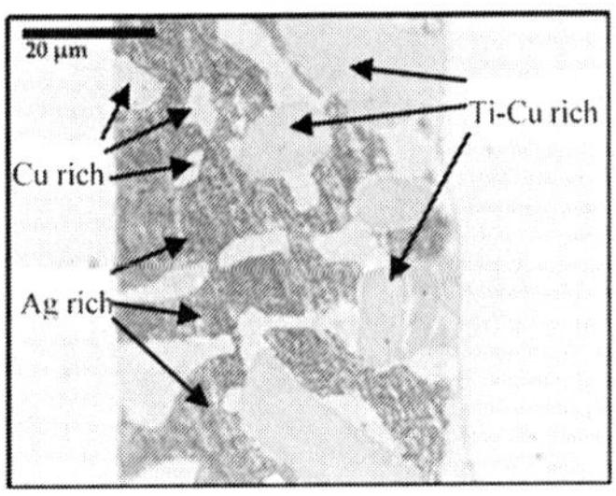


| Phase | Cu | Ag | Ti | In |
|---|---|---|---|---|
| Cu Rich | 90.18 | 5.86 | 2.09 | 1.88 |
| Ag Rich | 13.29 | 75.26 | 0 | 11.45 |
| Cu-Ti Rich | 77.27 | 0.84 | 21.21 | 0.67 |

(b)

**Figure 4.** Optical image of a Cu-Ag-Ti braze (**a**) and Cu-Ag-In-Ti braze (**b**) with labeled markers indicating the microconstituent phases present. Tables to the right of each image present average atomic % for each phase in (**a**) and (**b**), respectively, calculated from EDS data.

For the Cu-Ag-Ti braze alloy, two primary phases, composed primarily of either copper or silver, dominate the microstructure of the braze in agreement with previous findings [8]. This bead-like microstructure develops with overall low solubility of titanium within the Cu-Ag-Ti braze. Titanium was not detected in the eutectic phases, but was instead found along the interface between the braze and parent stainless steel metal [8]. Additionally, some titanium diffused into the neighboring stainless steel. This was also observed in previous research, and is attributed to the lack of both titanium solubility within the braze and miscibility within the adjacent stainless steel, leading to near-complete segregation [8,23]. In contrast, three distinct phases were observed in the Cu-Ag-In-Ti brazed joint. Also, segregation of the titanium to the interface was suppressed. Ti content varied in and near regions where a distinct Cu-Ti phase was observed (Figure 4b) and suggested incomplete phase transformation. The Cu-Ti rich phase present was likely the $Cu_4Ti$ intermetallic phase [23,24].

## 3.2. Corrosion Behavior of Brazed Joints

### 3.2.1. Electrochemical Testing of Braze Materials

Representative polarization curves from Cu-Ag-Ti (blue trace) and Cu-Ag-In-Ti (green trace) braze alloys are presented inFigure 5. Polarization data obtained from a bare 316L coupon (red trace) is also included in Figure 5 for comparison. The two different braze alloys displayed nearly identical polarization behavior. The braze alloys exhibited normal Tafel behavior indicative of activation polarization control with an corrosion rate, $i_{corr}$ (at open circuit) of approximately 1 × $10^{-5}$Amps/cm$^2$, approximately two orders of magnitude greater than the $i_{corr}$ of passive 316L. Moreover, the $E_{corr}$ (corrosion potential) of the braze alloys was approximately 50 mV less than that of 316L, −0.16 V *vs.* SCE.

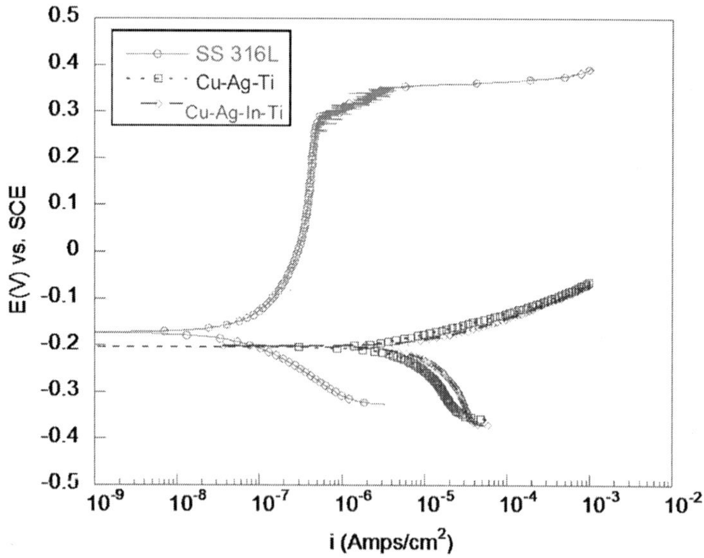


**Figure 5.** Potentiodynamic polarization scans conducted on brazes (blue and green traces) and a 316L stainless steel sample (red trace) in 0.6 M NaCl. The scan rate for all testing was 0.166 mV/s.

To determine any effect the thermal braze cycle might have on the inherent corrosion behavior of 316L, polarization curves were conducted on bare 316L samples that had been fired according to the braze cycle mentioned in the experimental section. When compared to the unfired stainless steel, the fired stainless steel sample displayed a much smaller passive region with pitting occurring at approximately 0.05 V *vs.* SCE (Figure 6). The pitting potential, or potential at which pitting corrosion occurs, was seen when a sharp increase in current occurred upon increasing potential. The pitting potential of the fired sample was significantly lower than the unfired 316L sample, likely due to grain boundary sensitization [25]. The effect on localized corrosion behavior could be expected since the sample was held in the sensitization range for approximately 35 min. Optical microscopy of the pitted surface (not shown here) revealed grain boundary attack had occurred at areas adjacent to pits on the fired sample. In contrast, pits on the unfired sample did not show evidence of preferential grain boundary attack near the pit openings, confirming some degree of sensitization on the 316L sample as a result of the thermal braze cycle.

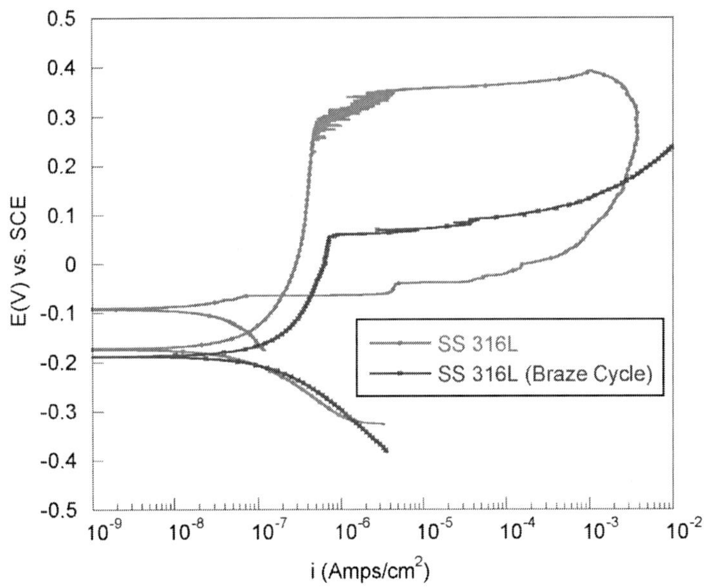


**Figure 6.** Potentiodynamic polarization scans conducted on 316L stainless steel subjected to a thermal brazing cycle (blue curve) compared to unfired 316L (red curve). Testing was conducted in 0.6 M NaCl with a scan rate 0.166 mV/s.

### 3.2.2. Exposure Testing at Open Circuit Potential

Exposure testing was conducted to observe macroscopic corrosion propagation behavior of the braze alloy and stainless steel at the free corrosion potential under natural, galvanic coupling conditions. Circular braze foil coupons were fabricated such that the foil occupied approximately 80% of the exposed surface area of the polished face of the 316L disk. Following firing, the coated samples and an untreated 316L control sample were immersed in 0.6 M NaCl solution and the samples were observed periodically to monitor corrosion progression, as shown in Figure 7.

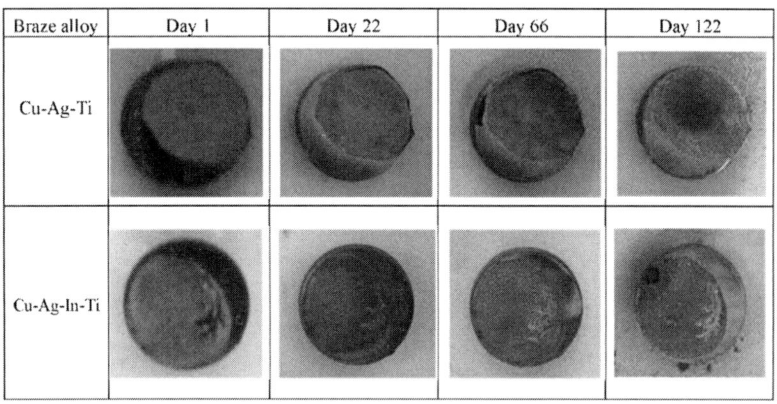


**Figure 7.** Time lapse photographs of corrosion propagation during long term exposure testing in 0.6 M NaCl. Cu-Ag-Ti (**top**) and Cu-Ag-In-Ti (**bottom**) braze alloy foil disks were used to coat ~80% of the exposed face area on 316L stainless steel samples. The stainless steel disk diameter was 16 mm for both samples.

The 316L control sample (not shown) did not display evidence of depassivation or production of any significant corrosion products for the duration of testing (166 days, or ~24 weeks). However, after only 22 days of immersion both of the braze foil coated samples showed evidence of macroscale corrosion damage (Figure 7). Visual observation revealed that corrosion was limited to the braze foil surface, while the remaining uncoated area of the 316L sample appeared unaffected. While the extent of the corrosion damage was not quantified, the presence of blue-green corrosion reaction products covering both the Cu-Ag-Ti and Cu-Ag-In-Ti foil surfaces suggested preferential corrosion of the Cu-rich phase present in each braze alloy [14]. Moreover, both samples also exhibited regions of dark red corrosion products deposited within the foil-coated region toward the end of the test duration. It is suspected these corrosion products were iron oxide-hydroxides generated from the underlying stainless steel which had become exposed following corrosion perforation of the braze foil coating.

## 3.3. SKPFM Measurements

SKPFM was used to measure Volta potential differences (VPD) among the various phases present on the surface of the brazed region of the stainless steel joint. SKPFM measurements were obtained prior to SEM/EDS characterization of the same regions on both braze alloy joints to avoid any effect of electron irradiation on surface potential measurements [22,26,27]. Potential maps acquired from SKPFM coupled with composition maps obtained from SEM/EDS clearly showed that the observed variations in potential correlated with changes in composition within the braze regions (Figure 8 andFigure 9).

From the SKPFM VPD data, there is a clear difference in potential values between the two primary phases for the Cu-Ag-Ti braze sample, with the Ag-rich phase as the brighter or more noble area, and the Cu-rich phase darker (more active potential). The Volta potential observed for the stainless steel outside the braze region was more negative than either of the braze phases (Figure 8b). These differences are expected based solely on composition, and the difference promotes dissimilar metal, galvanic couple driven corrosion within the braze alloy. Similar results were obtained on the Cu-Ag-In-Ti sample, with the most noble phase being the Ag-rich phase. The most active braze phase is the Cu-Ti rich phase, while the Volta potential of the Cu-rich phase falls in between the two. The surrounding stainless steel was similar to the Ti-Cu rich phase (Figure 9).

# 4. DISCUSSION

Research on braze reliability has largely focused on evaluating mechanical integrity of the joint [11,12,13,14]. Accordingly, corrosion behavior of brazes has received little attention in the open literature. The limited studies available focusing on stainless steel have considered different Ag-based braze compositions than studied here, but the systems behaved similarly and it was proposed that galvanic cells developed at the braze-stainless steel interface promoting corrosion driven by the electrically connected dissimilar metals [10,15,16,17,18,28]. However, in these studies, local galvanic couple corrosion was only postulated from bulk corrosion observations, it was not directly confirmed as in the present work with SKPFM. Other studies observed that partitioning of more noble alloying elements (preferred cathode sites) within the braze alloy occurred, causing depletion in the surrounding matrix and likely lead to preferential corrosion in those regions. In the systems investigated herein, similar corrosion behavior was confirmed [15,16,17]. Visual observation of corrosion propagation on braze alloy-stainless steel couples (Figure 7) showed that active corrosion rapidly initiated on the braze alloy first, with the surrounding uncoated stainless steel unaffected. The polarization curves presented in Figure 5 also suggest that the more anodic braze alloy will be preferentially attacked at an increased corrosion rate when galvanically connected to the more noble stainless steel cathode. Stainless steel had an open circuit potential (OCP) approximately 50 mV more noble that either of the braze alloys alone. On a macroscale, the stainless steel acts as cathode when galvanically coupled to the more anodic braze alloys, driving preferential corrosion of the braze alloy. This galvanic couple situation is particularly detrimental for joint integrity because of the typically large cathode (stainless steel) to anode (braze alloy) area ratio. At free corrosion conditions the anodic reaction occurring on the braze alloy is polarized towards the stainless steel, further accelerating the anodic reaction rate of the braze alloy. Because of the large cathode to anode area ratio there is ample cathode area available to support the increased dissolution rate of the braze, leading to eventual loss of joint integrity.

Following initiation, during corrosion propagation, the stainless steel also becomes susceptible to corrosion degradation near the braze interface, which is a further detriment to joint integrity. Corrosion attack on a 316L coupon with a foil braze coating on the surface showed that the copper phase was preferentially attacked, and the presence of the braze alloy served as a crevice former under which localized corrosion was able to propagate into the stainless steel (Figure 10). In Figure 10, the samples were exposed for 7 days to 0.6 M NaCl at OCP followed by a potentiodynamic scan from open circuit to 0.5 V*vs.* SCE, prior to cleaning and cross sectioning. The left image (Figure 10) was obtained from a cross section that bisected a corrosion pit which grew and propagated underneath the braze alloy coating. Eventual corrosion attack of the stainless steel in a brazed sample was also observed during free corrosion conditions when immersed in 0.6 M NaCl solution. As seen inFigure 7, the generation of voluminous dark red corrosion products indicated that braze alloy corrosion likely generated a sufficiently aggressive local chemical environment to depassivate the underlying stainless steel. While the uncoated regions of the stainless steel coupon did not display evidence of corrosion damage, it is evident that the galvanically driven braze alloy corrosion led to subsequent corrosion attack of the joined stainless steel at the interface for both braze alloys considered.

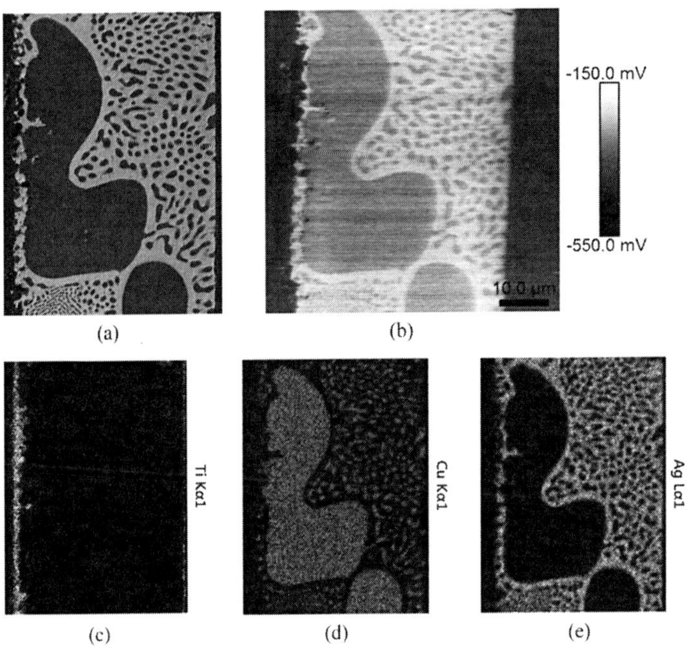


**Figure 8.** Secondary electron SEM image of Cu-Ag-Ti sample (**a**) followed by corresponding Scanning Kelvin Probe Force Microscopy (SKPFM) surface potential image (**b**). EDS elemental maps of the identical region for: Titanium (**c**); Copper (**d**); and Silver (**e**) are shown.

For the Cu-Ag-Ti braze alloy, corrosion caused by segregation of Ti to the braze-stainless steel interface (Figure 8) could also promote loss of joint coherency. The image on the right in Figure 10 shows preferential corrosion attack on the Cu-rich phase of the braze alloy and an area of corrosion damage along the 316L and Cu-Ag-Ti coating interface. The behavior of Ti was similar to that observed in literature [2] and can be attributed to the strong interfacial reactions between the Cu and Ti found in the braze [5]. The addition of In in the Cu-Ag-In-Ti samples did not support Ti segregation to the joint interface, and hence would be expected to be beneficial to joint integrity.

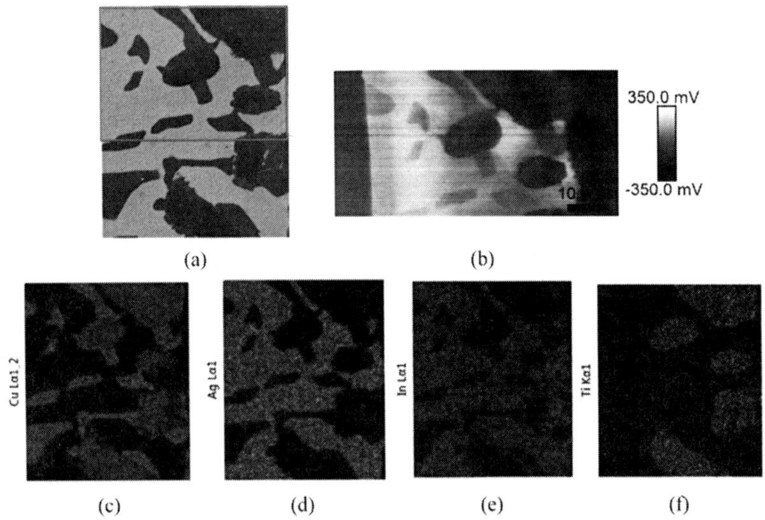


**Figure 9.** Secondary electron SEM image of Cu-Ag-In-Ti sample with the red box indicating where SKPFM was performed (**a**). SKPFM surface potential image (**b**) and EDS elemental maps of the identical region for: Copper (**c**); Silver (**d**); Indium (**e**); and Titanium (**f**) are shown.

In addition to the macroscale galvanic couple between braze alloys and stainless steel, both braze alloys formed multiple distinct phases following thermal treatment, resulting in microgalvanic cells between phases during active corrosion conditions. Open circuit potentials obtained from potentiodynamic polarization testing of both braze alloys were intermediate to those listed in the galvanic series for seawater for pure copper and pure silver [15,29]. Based on the galvanic series, the measured OCP of the bulk two phase (copper-rich and silver-rich) braze structure is in agreement with what would be expected from mixed potential theory. Within the braze alloy the expected electrode potential difference implies that when corrosion occurs, the Cu-rich phase of the braze alloy will be anodic to the more noble Ag-rich phase and suffer from preferential microgalvanic attack, even in the absence of a stainless steel couple.

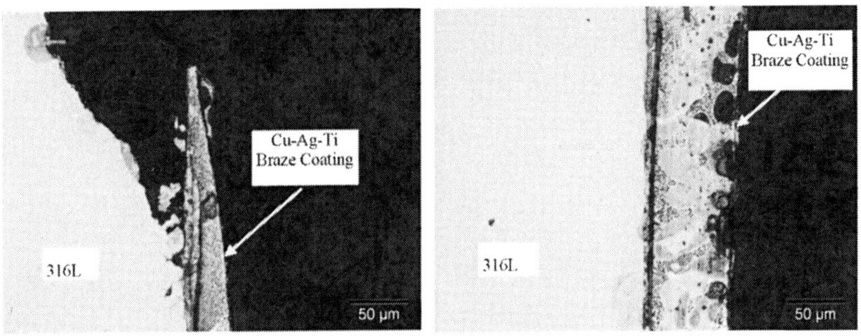


**Figure 10.** Cross sectioned 316L stainless steel sample with a Cu-Ag-Ti braze alloy coating following exposure for seven days in 0.6 M NaCl followed by a potentiodynamic scan.

The nature of microgalvanic corrosion within the braze alloys was further investigated with SKPFM. The goal of using SKPFM was to measure Volta potential differences between phases within the microstructure to determine relative nobility of the individual phases. Schmutz and Frankel, along with Leblanc, have established a direct correlation between Volta potentials measured in air and respective solution potentials of metals [26,27,30]. Importantly, the VPD observed was an effective indicator of how corrosion developed due to microstructural features. Subsequent to these findings, several research groups [31,32,33,34,35,36,37,38,39] have been able to verify that SKPFM is both reliable and effective in characterizing various alloy systems to accurately explain corrosion initiation behavior.

Galvanic corrosion on the microscale strongly influenced corrosion within the braze alloys. As determined via SKPFM, within the Cu-Ag-Ti braze, the Ag-rich phase was the most noble, followed by the Cu-rich phase, and finally stainless steel had the lowest overall potential (see Figure 11a, Table 3). For the Cu-Ag-In-Ti braze, the Ag-rich phase is the most noble, followed by the Cu-rich phase, then the Ti-Cu rich phase, as in Figure 12a,b and Table 4. The approximate average Volta potentials differences between phases in both braze alloys are presented in Table 3 and Table 4 with corresponding images of the SKPFM images of the brazes in Figure 11 and Figure 12, respectively. The SKPFM data presented in Figure 11 andFigure 12 are from identical areas characterized with SEM/EDS (Figure 8 and Figure 9). The co-localization of SKPFM and SEM/EDS techniques at the identical area provides direct evidence of the influence of local composition on Volta potential. Hence, in addition to the galvanic couple between the braze and stainless steel, variations in composition between phases provided the basis for microgalvanic corrosion within the braze alloys studied.

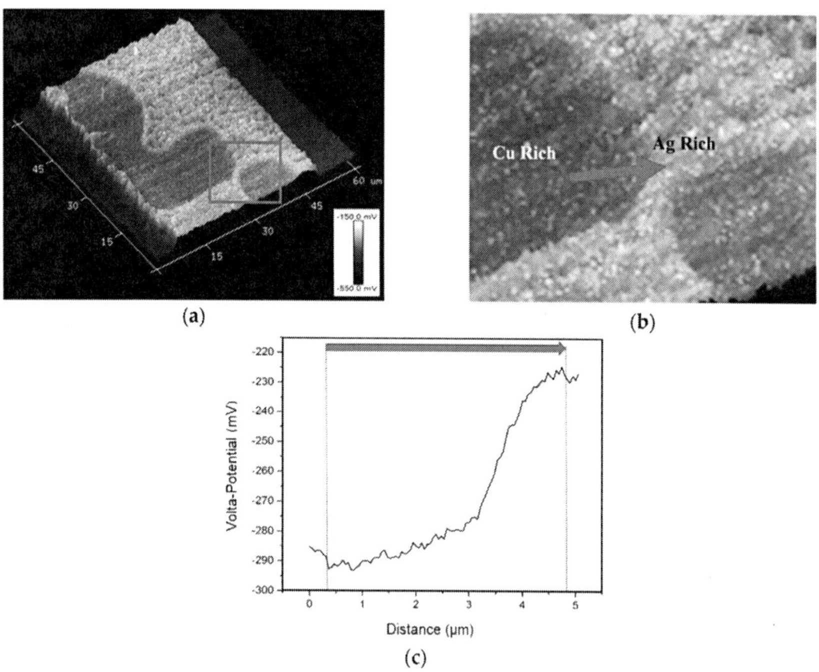


**Figure 11.** (**a**) Three dimensional (3-D) SKPFM Volta potential image of Cu-Ag-Ti sample with a box indicating area of scan; (**b**) Zoomed in image with arrow indicating where scan was taken; and (**c**) Graph of potential values obtained from scan as a function of distance with indicating arrow.

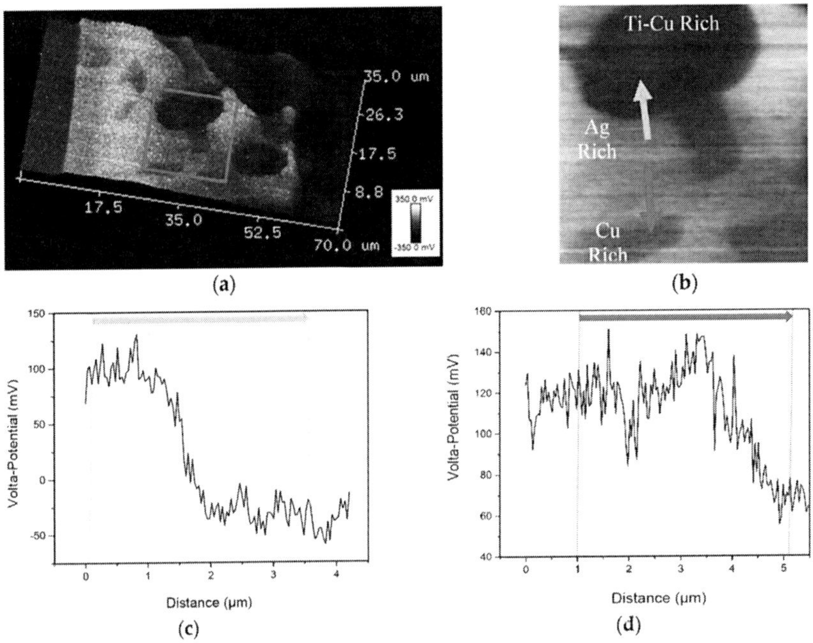


**Figure 12.** (**a**) 3-D SKPFM potential image of Cu-Ag-In-Ti sample with a box indicating area of scan. (**b**) Zoomed in image with arrows indicating where scan was taken. Graph of potential values obtained from scan as a function of distance with indicating arrow: (**c**) Blue arrow (Ag-rich phase to Ti-Cu-rich phase); and (**d**) Red arrow (Ag-rich phase to Cu-rich phase).

**Table 3.** Relative Volta potential difference (VPD) values in mV of phases within the Cu-Ag-Ti brazed joint. The first phase/metal is the more positive of the two.

| Microgalvanic Couple (Cathode–Anode) | ΔVPD (±30 mV) |
|---|---|
| Ag Rich–Cu Rich | 59 |
| Cu Rich–Stainless steel | 94 |

Interestingly, SKPFM results showed the stainless steel surface had an average Volta potential that was less noble than any of the braze phases measured (Figure 11 and Figure 12). This is contrary to the known and observed corrosion behavior, in that corrosion preferentially occurs within the braze alloy, with the coupled stainless steel acting as a cathode in the galvanic couple (Figure 5 and Figure 7). *This finding is important because it highlights an instance where caution must be exercised when inferring electrochemical behavior from Volta potential measurements.* In the case of stainless steel, the presence of a protective passive oxide layer makes it an ineffective anode, and the resulting solution potential is more noble than the braze material. Hence, it is essential to verify expected relative nobility obtained from SKPFM measurements with the observed corrosion and electrochemical behavior in solution, as done here.

**Table 4.** Relative VPD in mV of the Cu-Ag-In-Ti sample. The first phase/metal is the more positive of the two.

| Microgalvanic Couple (Cathode–Anode) | ΔVPD (±30 mV) |
|---|---|
| Ag Rich–Cu Rich | 52 |
| Cu Rich–(Ti-Cu) Rich | 97 |
| (Ti-Cu) Rich–Stainless Steel | 53 |

# 5. CONCLUSIONS

The composition, phase separation, surface potential differences, and corrosion behavior of Cu-Ag braze alloys (Cu-Ag-Ti and Cu-Ag-In-Ti) and the joined material (316L) were investigated and correlated. SKPFM measurements provided new insight to the origins of microgalvanic corrosion within the brazed region and confirmed the manner that corrosion develops, which previously had only been postulated to explain bulk corrosion observations of brazed joints. Significant findings include:

1) Co-localized SKPFM and SEM/EDS provided evidence of phase separation within the braze regions that resulted in surface potential differences. Moreover, the measured surface potentials correlated with the observed microgalvanic corrosion behavior, thereby highlighting the utility of combining these methods for the future study, prediction, and prevention of microgalvanic corrosion.

2) Microgalvanic cells were confirmed via SKPFM VPD values. The two phases present in the Cu-Ag-Ti braze alloy samples differed by ~60

mV in surface potential, while the Cu-Ag-In-Ti samples exhibited a range in surface potential differences up to ~250 mV across three phases.

3) Electrochemical testing on the individual materials was used to verify expected galvanic behavior. Stainless steel exhibited passive behavior and had an OCP that was noble to either of the braze alloys, which exhibited active corrosion behavior and $i_{corr}$ approximately two orders of magnitude larger than stainless steel. The role of stainless steel in the direction of the galvanic couple was in conflict with that expected from SKPFM measurements and provided an exceptional case where SKPFM Volta potential did not correlate with solution potential.

4) Exposure testing of stainless steel samples coated with each braze alloy were conducted at OCP in 0.6 M NaCl. It was found the braze alloy undergoes active corrosion that induces accelerated attack on the underlying 316L material. Additionally, it was seen that the Cu-rich phase of the braze alloy underwent preferential attack compared to the Ag-rich phase and corrosion propagation in the braze alloy was aggressive enough to depassivate the adjacent 316L stainless steel.

5) The thermal brazing cycle caused sensitization of the 316L parent material and resulted in grain boundary attack and a pitting potential that was approximately 200 mV more negative than the unfired 316L.

## ACKNOWLEDGMENTS

Fabrication of samples was supported by National Science Foundation contract no. OISE-0709664. The generous support of the Boise State Surface Science Laboratory (SSL), the Center for Advanced Energy Studies (CAES), and Central Metallurgical Research & Development Institute, Egypt are also acknowledged. The authors thank Jasen B. Nielsen for assistance with SKPFM sample preparation.

## AUTHOR CONTRIBUTIONS

Michael F. Hurley, Paul H. Davis and Darryl P. Butt conceived and designed the experiments; Corey M. Efaw, Armen Kvryan and Michael F. Hurley performed the experiments; Kari Livingston, Armen Kvryan, Paul H. Davis and Corey M. Efaw analyzed the data; Kyle Knori, Brian J. Jaques, Darryl P. Butt and Paul H. Davis contributed reagents, materials and analysis tools. All authors contributed to writing the paper.

## REFERENCES


1. Jacobson, D.M.; Humpston, G. *Principles of Brazing*; ASM International: Materials Park, NY, USA, 2005.

2. Dev, S.C.; Singh, I.; Basu, D.K.; Bhattamishra, A.K.; Sivaramakrishnan, C.S. Corrosion behaviour of silver brazing alloys in different environments. *Anti-Corros. Methods Mater.* **1997**, *44*, 260–264.

3. Dimatteo, A.; Lovicu, G.F.; Desanctis, M.; Valentini, R.; Santarelli, S.; Cognetta, F. Compatibility of nickel and silver-based brazing alloys with E85 fuel blends. *Mater. Corros.* **2015**, *66*, 158–168.

4. Shafrin, E.G.; Zisman, W.A. Constitutive relations in the wetting of low energy surfaces and the theory of the retraction method of preparing monolayers. *J. Phys. Chem.* **1960**, *64*, 519–524.

5. Dezellus, O.; Eustathopoulos, N. Fundamental issues of reactive wetting by liquid metals. *J. Mater. Sci.* **2010**, *45*, 4256–4264.

6. Paiva, O.C.; Barbosa, M.A. Microstructure, mechanical properties and chemical degradation of brazed AISI 316 stainless steel/alumina systems. *Mater. Sci. Eng.* **2008**, *480*, 306–315.

7. Kozlova, O.; Voytovych, R.; Devismes, M.F.; Eustathopoulos, N. Wetting and brazing of stainless steels by copper-silver eutectic. *Mater. Sci. Eng.* **2008**, *495*, 96–101.

8. Abed, A.; Jalham, I.S.; Hendry, A. Wetting and reaction between β′-sialon, stainless steel and Cu-Ag brazing alloys containing Ti. *J. Eur. Ceram. Soc.* **2001**, *21*, 283–290.

9. Nicholas, M.G.; Valentine, T.M.; Waite, M.J. The wetting of alumina by copper alloyed with titanium and other elements. *J. Mater. Sci.* **1980**, *15*, 2197–2206.

10. Takemoto, T.; Okamoto, I. Effect of composition on the corrosion behavior of stainless-steels brazed with silver-base filler metals. *Weld. J.* **1984**, *63*, S300–S307.

11. Lee, J.G.; Hong, S.J.; Lee, M.K.; Rhee, C.K. High strength bonding of titanium to stainless steel using an Ag interlayer.*J. Nucl. Mater.* **2009**, *395*, 145–149.

12. Liu, C.C.; Ou, C.L.; Shiue, R.K. The microstructural observation and wettability study of brazing Ti-6Al-4V and 304 stainless steel using three braze alloys. *J. Mater. Sci.* **2002**, *37*, 2225–2235.

13. Singh, M.; Shpargel, T.P.; Asthana, R. Brazing of stainless steel to yttria-stabilized zirconia using gold-based brazes for solid oxide fuel cell applications. *Int. J. Appl. Ceram. Technol.* **2007**, *4*, 119–133.

14. Singh, M.; Shpargel, T.P.; Asthana, R. Brazing of yttria-stabilized zirconia (YSZ) to stainless steel using Cu, Ag, and Ti-based brazes. *J. Mater. Sci.* **2008**, *43*, 23–32.

15. Anastasio, S.; James, J.; Fitz-Gerald, J.; Scullly, J.R. Corrosion of superaustenitic stainless steel N08367 brazed with a nickel-chromium-silicon-phosphorous alloy: Electrochemical corrosion behavior of isolated and combined materials.*Corrosion* **2009**, *65*, 388–403.

16. James, J.P.; Bocher, F.; Scully, J.R. Effect of braze clearance on localized corrosion of a superaustenitic stainless steel brazed with a Ni-based alloy (Ni-22Cr-6.3Si-3.8P). *Corrosion* **2009**, *65*, 511–526.

17. Sorensen, N.R. The environmentally assisted failure of cusil in mattssons solution. *Corrosion* **1986**, *42*, 299–306.

18. Lee, M.K.; Park, J.J.; Lee, J.G.; Rhee, C.K. Phase-dependent corrosion of titanium-to-stainless steel joints brazed by Ag-Cu eutectic alloy filler and Ag interlayer. *J. Nucl. Mater.* **2013**, *439*, 168–173.

19. Chiu, L.H.; Wu, C.H.; Chang, H. Galvanic corrosion on vacuum-brazed UNS S31803 duplex stainless steel using Ni-Cr-Fe-P alloy filler metals. *Corrosion* **2007**, *63*, 127–134.

20. Guillaumin, V.; Schmutz, P.; Frankel, G.S. Characterization of corrosion interfaces by the scanning kelvin probe force microscopy technique. *Electrochem. Soc.* **2001**, *148*, 163–173.

21. Schmutz, P.; Frankel, G.S. Corrosion study of AA2024-T3 by scanning kelvin probe force microscopy and *in situ*atomic force microscopy scratching. *J. Electrochem. Soc.* **1998**, *145*, 2295–2306.

22. Hurley, M.F.; Efaw, C.M.; Davis, P.H.; Croteau, J.R.; Graugnard, E.; Birbilis, N. Volta potentials measured by scanning kelvin probe force microscopy as relevant to corrosion of magnesium alloys. *Corrosion* **2015**, *71*, 160–170.

23. Yue, X.; He, P.; Feng, J.C.; Zhang, J.H.; Zhu, F.Q. Microstructure and interfacial reactions of vacuum brazing titanium alloy to stainless steel using an agcuti filler metal. *MTL Mater. Charact.* **2008**, *59*, 1721–1727.

24. Elrefaey, A.; Tillmann, W. Effect of brazing parameters on microstructure and mechanical properties of titanium joints. *J. Mater. Process. Technol.* **2009**, *209*, 4842–4849.

25. Jones, D. *Principles and Prevention of Corrosion*, 2nd ed.; Prentice Hall: Upper Saddle River, NJ, USA, 1996.

26. Schmutz, P.; Frankel, G.S. Characterization of AA2024-T3 by scanning kelvin probe force microscopy. *J. Electrochem. Soc.* **1998**, *145*, 2285–2295.

27. Schmutz, P.; Frankel, G.S. Influence of dichromate ions on corrosion of pure aluminum and AA2024-T3 in NaCl solution studied by AFM scratching. *J. Electrochem. Soc.* **1999**, *146*, 4461–4472.

28. Chiu, L.H.; Hsieh, W.C.; Ling, Y.C. Effect of vacuum brazing on corrosion resistance of UNS S31803 and UNS S31200 duplex stainless steels. *Corrosion* **2002**, *58*, 797–803.

29. Forman, C.M.; Verchot, E.A. *Practical Galvanic Series*; Report No. RS-TR-67-11; Army Missile Command: Redstone Arsenal, AL, USA, 1967.

30. Leblanc, P.; Frankel, G.S. A study of corrosion and pitting initiation of AA2024-T3 using atomic force microscopy. *J. Electrochem. Soc.* **2002**, *149*, B239–B247.

31. Jia, J.X.; Atrens, A.; Song, G.; Muster, T.H. Simulation of galvanic corrosion of magnesium coupled to a steel fastener in NaCl solution. *Mater. Corros.* **2005**, *56*, 468–474.

32. De Wit, J.H.W. Local potential measurements with the SKPFM on aluminium alloys. *Electrochim. Acta* **2004**, *49*, 2841–2850.

33. Campestrini, P.; van Westing, E.P.M.; van Rooijen, H.W.; de Wit, J.H.W. Relation between microstructural aspects of AA2024 and its corrosion behaviour investigated using AFM scanning potential technique. *Corros. Sci.* **2000**, *42*, 1853–1861.

34. Lacroix, L.; Ressier, L.; Blanc, C.; Mankowski, G. Combination of AFM, SKPFM, and SIMS to study the corrosion behavior of S-phase particles in AA2024-T351. *J. Electrochem. Soc.* **2008**, *155*, C131–C137.

35. Femenia, M.; Canalias, C.; Pan, J.; Leygraf, C. Scanning kelvin probe force microscopy and magnetic force microscopy for characterization of duplex stainless steels. *J. Electrochem. Soc.* **2003**, *150*, B274–B281.

36. Sathirachinda, N.; Pettersson, R.; Pan, J.S. Depletion effects at phase boundaries in 2205 duplex stainless steel characterized with SKPFM and TEM/EDS. *Corros. Sci.* **2009**, *51*, 1850–1860.

37. Sathirachinda, N.; Pettersson, R.; Wessman, S.; Kivisakk, U.; Pan, J.S. Scanning kelvin probe force microscopy study of chromium nitrides in 2507 super duplex stainless steel-implications and limitations. *Electrochim. Acta* **2011**, *56*, 1792–1798.

38. Mato, S.; Alcala, G.; Woodcock, T.G.; Gebert, A.; Eckert, J.; Schultz, L. Corrosion behaviour of a Ti-base nanostructure-dendrite composite. *Electrochim. Acta* **2005**, *50*, 2461–2467.

39. Afshar, F.N.; de Wit, J.H.W.; Terryn, H.; Mol, J.M.C. Scanning kelvin probe force microscopy as a means of predicting the electrochemical characteristics of the surface of a modified AA4XXX/AA3XXX (Al alloys) brazing sheet.*Electrochim. Acta* **2013**, *88*, 330–339

9

# HOMOGENIZATION HEAT TREATMENT TO REDUCE THE FAILURE OF HEAT RESISTANT STEEL CASTINGS

**Mohammad Hosein Bina**[1]

[1] Department of Advanced Materials and New Energy, Iranian Research Organization for Science and Technology, Tehran, Iran

## 1. INTRODUCTION

In this chapter, influence of the homogenization heat treatment on the failure of heat resistant steel castings is studied. First, the cast stainless steels and effective factors on the failure of these steels are described. One of the most important factors is sigma phase embrittlement which is studied, in detail. Finally, the effect of homogenization heat treatment on dissolution of carbides and reduction of sigma-phase and failure is discussed. The present chapter is a compilation of my experience in industry and other studies about fracture of continuous annealing furnace rollers, prepared for use by practitioners and researchers. The chapter may also be useful for graduate students, researching failure.

## 2. CLASSIFICATION AND DESIGNATION OF CAST STAINLESS STEELS

Cast stainless steels are usually classified as either corrosion resistant steel castings (which are used in aqueous environments below 650°C) or heat resistant steel castings (which are suitable for service temperatures above 650 °C). Cast stainless steels are most often specified on the basis of chemical composition using the designation system of the High Alloy Product Group of the Steel Founders Society of America [1].

## 2.1. Corrosion Resistant Steel Castings

The corrosion resistant steel castings are widely used in chemical processes and electricity equipment that need corrosion resistant in aqueous or liquid-vapor environments at the temperatures less than 315°C. The serviceability of cast corrosion-resistant steels depends greatly on the absence of carbon, and especially precipitated carbides, in the alloy microstructure. Therefore, cast corrosion-resistant alloys are generally low in carbon (usually lower than 0.20% and sometimes lower than 0.03%). All cast corrosion-resistant steels contain more than 11% chromium [1].

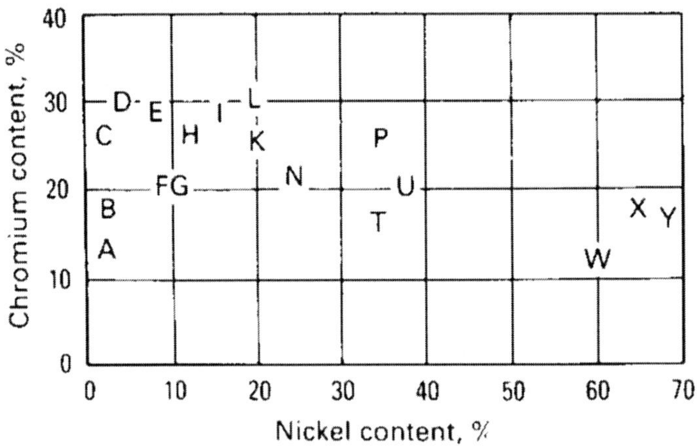


**Figure 1.** Chromium and nickel contents in ACI standard grades of heat- and corrosion-resistant steel castings [**1**].

## 2.2. Heat Resistant Steel Castings

Stainless steel castings are classified as heat resistant if they are capable of sustained operation while exposed, either continuously or intermittently, to operating temperatures that result in metal temperatures in excess of 650°C. Heat resistant steel castings resemble high-alloy corrosion resistant steels except for their higher carbon content, which imparts greater strength at elevated temperature [1]. The three principal categories of H-type cast steels, based on composition, are [1–4]:

a) Iron-chromium alloys containing 10 to 30% Cr and little or no nickel. These alloys have low strength at elevated temperatures and are useful mainly due to their resistance to oxidation. Use of these alloys is restricted to conditions, either oxidizing or reducing, that involve low static loads and uniform heating. Chromium content depends on anticipated service temperature [1].

b) Iron-chromium-nickel alloys contain more than 13% Cr and 7% Ni (always more chromium than nickel) [1]. These austenitic alloys are ordinarily used under oxidizing or reducing conditions similar to those withstood by the ferritic iron-chromium alloys, but in service they have greater strength and ductility than the straight chromium alloys. They are used, therefore, to withstand greater loads and moderate changes in temperature. These alloys also are used in the presence of oxidizing and reducing gases that are high in sulfur content.

c) Iron-nickel-chromium alloys contain more than 25% Ni and more than 10% Cr (always more nickel than chromium) [1]. These austenitic alloys are used for withstanding reduction as well as oxidizing atmospheres, except where sulfur content is appreciable. (In atmospheres containing 0.05% or more hydrogen sulfide, for example, iron-chromium-nickel alloys are recommended [1].) In contrast with iron-chromium-nickel alloys, iron-nickel-chromium alloys do not carburize rapidly or become brittle and do not take up nitrogen in nitriding atmospheres. These characteristics become enhanced as nickel content is increased, and in carburizing and nitriding atmospheres casting life increases with nickel content [1]. Austenitic iron-nickel-chromium alloys are used extensively under conditions of severe temperature fluctuations such as those encountered by fixtures used in quenching and by parts that are not heated uniformly or that are heated and cooled intermittently. In addition, these alloys have characteristics that make them suitable for electrical resistance heating elements.

# 3. MICROSTRUCTURE OF HEAT RESISTANT STEELS

The microstructure of a particular grade is primarily determined by composition. Chromium, molybdenum, and silicon promote the formation of ferrite (magnetic), while carbon, nickel, nitrogen, and manganese favor the formation of austenite (nonmagnetic). Chromium (a ferrite and martensite promoter), nickel, and carbon (austenite promoters) are particularly important in determining microstructure [1]. The heat resistant casting alloys (except HA-type) are contain high proportions of chromium and nickel that are generally austenitic and nonmagnetic. The austenite in the matrix provides useful high-temperature strength if it is adequately reinforced with particles of carbide and nitride. The austenite must contain no ferrite to reach maximum strength [5]. The microstructure of HH-type austenitic stainless steel castings (25Cr-12Ni) used in continuous annealing furnace rollers in un-worked condition is presented in Fig. 2. The ideal microstructure of HH alloy is austenitic, while, in Fig. 2, the initial microstructure consisted of austenite, ferrite, and also black carbide particles. The presence of ferrite was attributed to un-controlled solidification and therefore, occurrence segregation phenomenon in the sample.

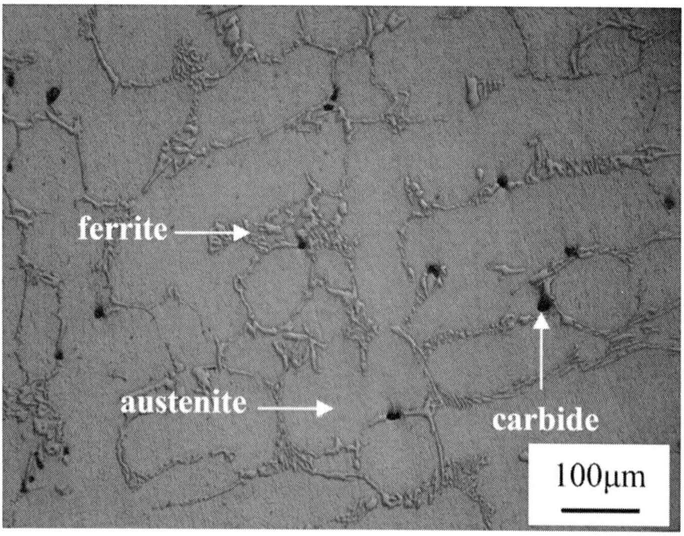


**Figure 2.** Microstructure of roller at un-worked condition (2% nital agent).

The presence of ferrite in microstructure may be beneficial or detrimental (especially at high temperatures), depending on the application. Ferrite can be beneficial in terms of weldability because fully austenitic stainless steels are susceptible to a weldability problem known as hot cracking, or microfissuring. But, at 540°C and above, the ferrite phase may transform to a complex iron-chromium-nickel-molybdenum intermetallic compound known as σ phase, which reduces toughness, corrosion resistance, and creep ductility [1]. The extent of the reduction in strength increases with time and temperature to about 815°C and may persist to 925°C [1].

# 4. SIGMA PHASE EMBRITTLEMENT

Several factors cause failure of heat resistant steel castings such as oxidation, sulfidation, carburization, creep, thermal fatigue, and sigma phase embrittlement. Intermetallics such as the sigma phase are important sources of failure in high-temperature materials.

The existence of σ phase in iron-chromium alloys was first detected in 1907 by the observation of a thermal arrest in cooling curves [6]. The first actual observation of σ in iron-chromium alloys was reported in 1927 [7]. The σ phase was identified by x-ray diffraction in 1927 [8] and in 1931 [9]. After the existence of σ was firmly established, numerous studies were conducted to define the compositions and temperatures over which σ could be formed. In general, σ forms with long-time exposure in the range of 565 to 980°C, although this range varies somewhat with composition and processing [1]. Sigma phase has a tetragonal crystal structure with 30 atoms per unit cell and a $c/a$ ratio of approximately 0.52 [10]. Sigma also forms in austenitic alloys. In fully austenitic alloys, σ forms from the austenite along grain boundaries. If δ-ferrite is present in the austenitic alloy, σ formation is more rapid and occurs in the δ-ferrite [1].

This phase is a unique combination of iron and chromium that produces a hard and brittle second phase. The presence of sigma phase not only results in harmful influence on the mechanical properties of the material, but also reduces its corrosion resistance by removing chromium and molybdenum from the austenitic matrix. A relatively small amount of sigma phase, when it is nearly continuous at a grain boundary, can lead to the very early failure of parts [11]. Fig. 3(a) shows a failed roller (HH-type) which was used for metal strip transfer in continuous annealing furnaces. An austenitic matrix and a network of sigma phase precipitates on austenitic grain boundaries in the failed roller is shown in Fig. 3(b). Formation of the sigma phase during prolonged heating at temperatures exceeding 650°C, results in a detrimental decrease in roller toughness at lower temperatures and during shut-down periods. Therefore, during shut-down, straightening or heating-up periods, the crack propagation along sigma phase at grain boundaries result in failure of rollers [12,13].

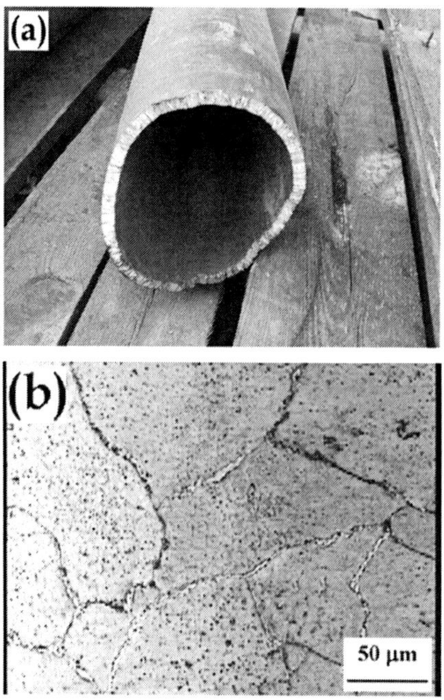


**Figure 3.** a) View of the failed furnace roller, (b) microstructure of austenitic matrix and network of sigma phase precipitation on grain boundary (electrolytic etching) [**13**].

All of the ferritic stabilizing elements promote σ formation [1]. In commercial alloys, silicon, even in small amounts, markedly accelerates the formation of σ. In general, all of the elements that stabilize ferrite promote σ formation. Molybdenum has an effect similar to that of silicon, while aluminum has a lesser influence. Increasing the chromium content, also favors σ formation. Small amounts of nickel and manganese increase the rate of σ formation, although large amounts, which stabilize austenite, retard σ formation. Carbon additions decrease σ formation by forming chromium carbides, thereby reducing the amount of chromium in solid solution [14,15]. Additions of tungsten, vanadium, titanium, and niobium also promote σ formation. As might be expected, σ forms more readily in ferritic than in austenitic stainless steels [14–17].

## 4.1. Formation Mechanism Of Sigma Phase In Heat Resistant Steels

Fig. 4 shows the formation mechanism of sigma phase in the failed roller. The presence of delta-ferrite in the austenitic matrix leads to accelerated sigma-phase formation. Generally, this phase nucleates inside delta-ferrite. Transformation of delta-ferrite can be described by two eutectoid reactions [1]:

$$\delta \rightarrow M_{23}C_6 + \gamma_2 \quad (1)$$

$$\delta \rightarrow \sigma + \gamma_2 \quad (2)$$

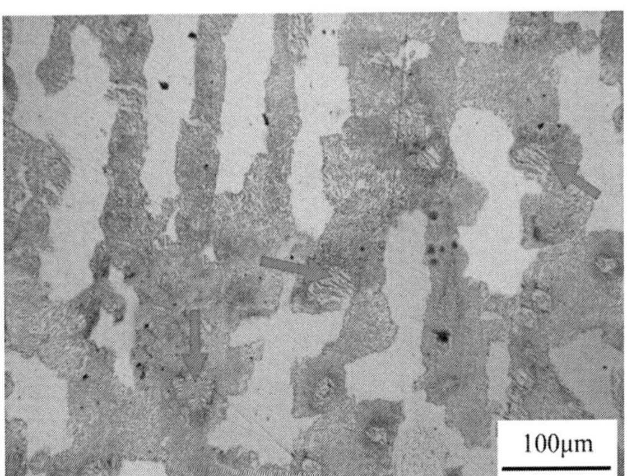


**Figure 4.** Formation of pearlite-like structure in the cross section of failed roller. The arrows show the large pearlite-like colonies.

By reaction (1), $M_{23}C_6$ carbides with lamellar morphology and also secondary austenite formed (Fig. 4). The mechanism of this reaction can be summarized as follows: first, the growth of carbide precipitates inside delta-ferrite lead to the formation of secondary austenite [18]. The chromium content near precipitates is high and results in increase in the content of chromium close to adjacent delta-ferrite. Therefore, the carbides growth develops inside delta-ferrite and a lamellar structure including carbides and secondary austenite is formed [19]. Finally, after completion of lamellar precipitation, the sigma-phase forms at foreside of the precipitates [19].

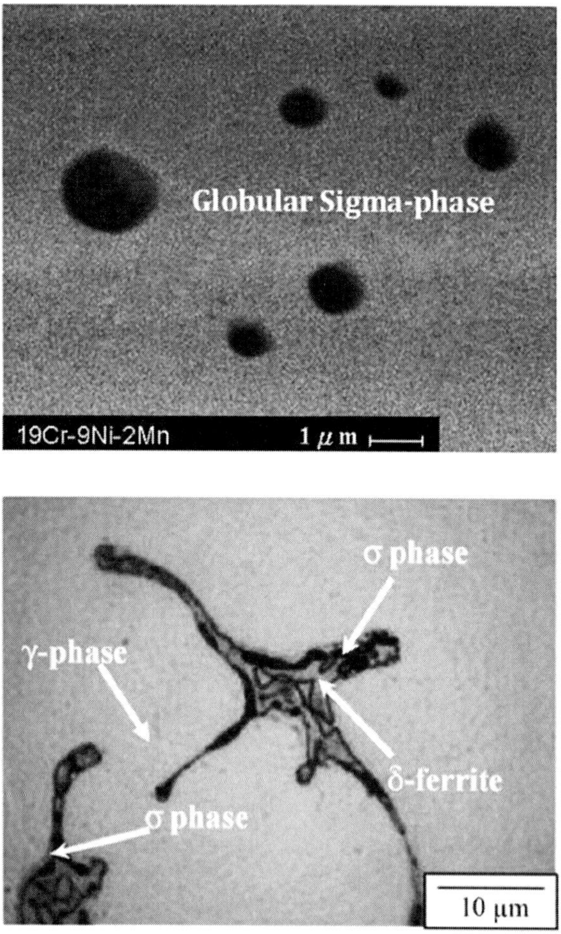


**Figure 5.** Various morphologies of sigma phase in a stainless steel [**20**].

## 4.2. Morphology Of Sigma Phase

The morphologies of sigma-phase can be classified into dendritic and globular structures [17,20,21].Fig. 5 shows the morphologies of sigma phase in a stainless steel (19Cr-9Ni-2Mn). The sigma-phase was formed by transformation of the delta-ferrite to the embrittling sigma-phase and secondary austenite due to unsuitable working conditions. Lin et al. [20] have pointed out that the dendrite-like sigma-phase is unstable, while the globular sigma-phase is a stable one. Furthermore, dendrite-like σ morphology was observed surrounding the δ-ferrite particles, which meant that the $\delta \rightarrow \sigma$ phase transformation occurred partially [20]. Gill et al. [21] also have proposed that the spheroid σ phase results from the instability of dendritic sigma-phase to any localized decrease in width.

## 4.3. Precipitation Sites Of Sigma Phase

The precipitation sites of sigma-phase consist of δ/γ interface boundary, triple conjunction, grain corner and cellular which are shown in Fig. 6 [22]. The initial precipitation sites are δ/γ interphase boundary because it has higher boundary energy and many defects are concentrated here. Therefore, the precipitation of sigma-phase takes place preferentially at δ/γ boundary, and then precipitates toward interior of delta-ferrite grain [19,22]. The other precipitation sites were concentrated at δ-ferrite because σ-phase preferred to precipitate at a higher Cr content region [20,22]. The precipitation at grain corner meant that sigma-phase was strongly concentrated and formed at the corner of delta-ferrite. The precipitation at triple conjunction meant the σ-phase precipitated surrounding δ-ferrite. The cellular shape precipitation presented the eutectoid decomposition from delta-ferrite into sigma and secondary austenite phases that can be observed clearly at 800°C [22].

## 4.4. Effect Of Sigma Phase On Mechanical Properties

One of the most affected mechanical properties of steels by formation of the sigma phase is impact energy. Effect of sigma phase on the impact energy of austenitic steel Fe-25Cr-20Ni is shown in Fig. 8[23]. By increasing the time of exposure at formation temperature range of sigma phase (760–870°C), toughness value decreases by 85%.

The influence of high-temperature exposure on the toughness of a low-interstitial 29Cr-4Mo ferritic stainless steel has been examined (Fig. 9) by Aggen et al. [24]. In this figure, the C-curve shows the time for embrittlement as a function of aging temperature. For σ formation, embrittlement was most rapid at about 775°C, whereas 475°C embrittlement was slower with a maximum rate at about 480°C. In general, for this alloy, the sigma-phase forms over the range of 595 to 925°C. Embrittlement is most pronounced when intergranular σ films form, resulting in intergranular tensile and impact fractures [1].

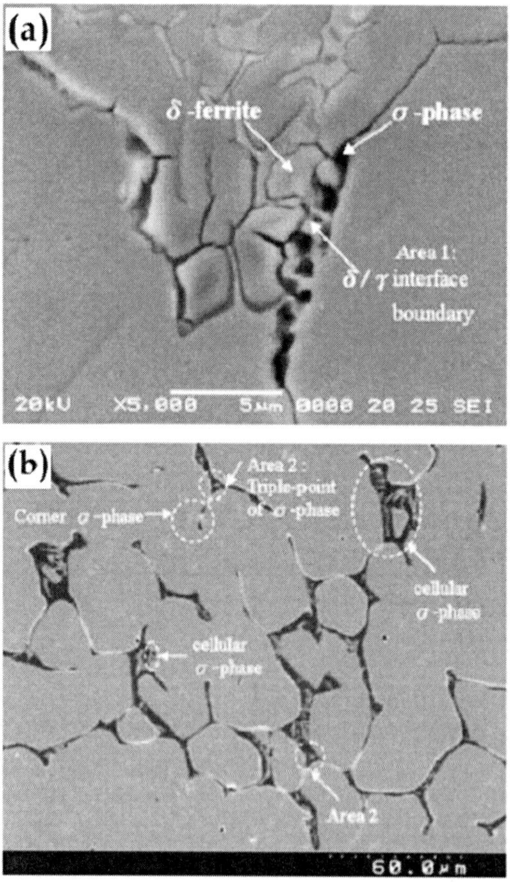


**Figure 6.** Precipitation of sigma-phase at: (a) δ/γ interface boundary, and (b) triple conjunction, grain corner and cellular [**22**].

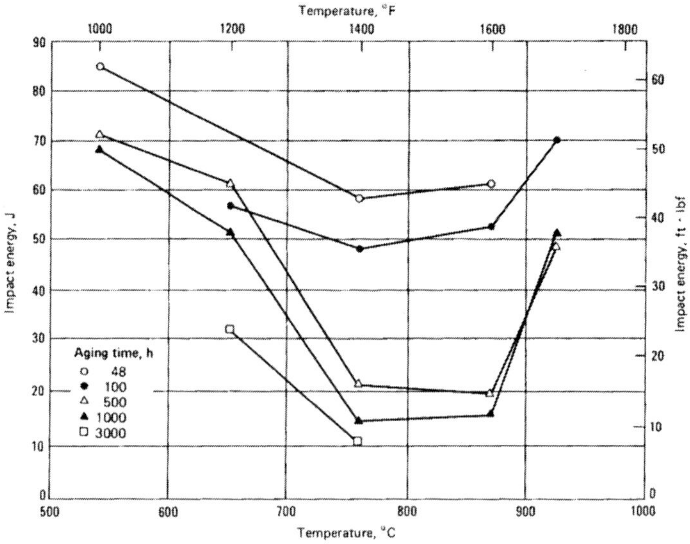


**Figure 7.** Effect of time and temperature of aging treatment on the impact energy of austenitic steel Fe-25Cr-20Ni [**23**].

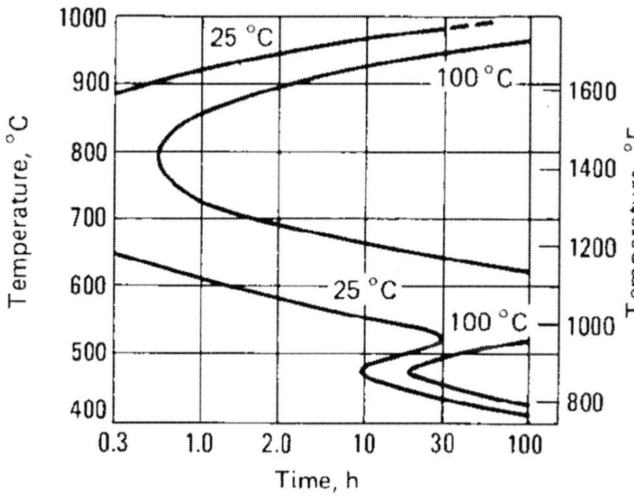


**Figure 8.** Time-temperature relationships to produce 25 and 100°C DBTTs for a 29Cr-4Mo ferritic stainless steel as a function of aging times that cover both the 475°C embrittlement range and the σ phase embrittlement range [**24**].

The hardness of sigma phase in Fe-Cr alloys is approximately 68 HRC [1]. Fig. 10 shows the microstructure of austenitic matrix and network of sigma phase precipitation on grain boundary and also indenter effects on the surface at several locations. It can be seen that because of brittleness and higher hardness of sigma phase with respect to the austenite matrix, this phase often fractures during indentation.

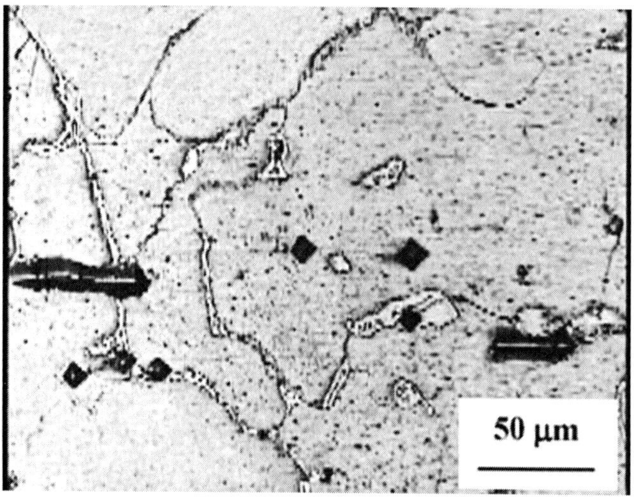


**Figure 9.** Microstructure of austenitic matrix and network of sigma phase precipitation on grain boundary. It can be seen that because of sigma is brittle phase, often fractures during indentation (indicated by the arrows) [**12**].

Therefore, a relatively small quantity of the σ phase, when it is nearly continuous at a grain boundary, can lead to very early failure of high-temperature parts [12].

## 5. HOMOGENIZATION HEAT TREATMENT

To eliminate segregation in the cast structure, it is frequently necessary to homogenize the part before usage to promote uniformity of chemical composition and microstructure. This treatment can also be used for complete dissolution of carbides and brittle and deleterious phases like sigma which are formed at operating conditions. In this method, steel is heated for a long enough period of time in order to complete the dissolution of carbides, and then the cooling starts at an appropriate rate to avoid formation of deleterious phases [12].

Microstructures of selected failed roller steel, before and after the homogenization heat treatment at various temperatures for 2 h are shown in Table 1. Formation of the brittle sigma phase and precipitation of carbides is clear in the microstructure of specimen HH14 when compared to the microstructure of specimen HH11. As a result of slow cooling in the furnace, all the desirable changes in the homogenized structure are reversed. In other words, the formation of carbide precipitates in grains, inhomogeneity of austenite as the matrix phase, and the formation of intermetallic brittle phases like sigma phase, which are all results of slow cooling at temperatures between 650 and 950°C, cause the toughness of the steel to decrease greatly. Since the formation of the sigma phase is a time consuming process, formation of this phase has been minimized by cooling in the air. The lowest amounts of sigma phase and carbides are in specimen HH11, which shows homogeneity of microstructure and a decrease in microscopic segregation. Therefore, for roller straightening and a good toughness to be obtained, homogenization heat treatment should be performed at 1100°C for at least 2 h followed by cooling in air [12].

Fig. 11 shows the precipitation sites of sigma-phase in the failed roller before and after the homogenization heat treatment at 1100 °C for 2 h followed by air cooling. Fig. 12 and Table 2 illustrate the chemical composition at three scanning points indicated in Fig. 11 (a), (b) and (c). As can be seen in Fig. 11, the points 1, 2, and 3 were austenite matrix, delta-ferrite, and sigma-phase, respectively.Table 2 shows that the chromium, molybdenum, and silicon content in the sigma-phase were higher than that of the austenite-phase. Generally, delta-ferrite and sigma-phase are Cr-rich. But, the weight percent of silicon and molybdenum in the delta-ferrite is higher than that of other phases (austenite and sigma). Silicon and molybdenum plays an important role in $\delta \rightarrow \sigma + \gamma 2$ phase transformation and acts as strong stabilizers for delta-ferrite [5,20].

**Table 1.** Microstructures of selected failed roller steel, before and after the homogenization heat treatment for 2 h.

| Specimen | Homogenization temperature (°C) | Cooling environment | Microstructures (etched electrolytically by 10 M KOH solution) | Microstructures (etched by Marble's reagent) |
|---|---|---|---|---|
| HH1 | - | - | 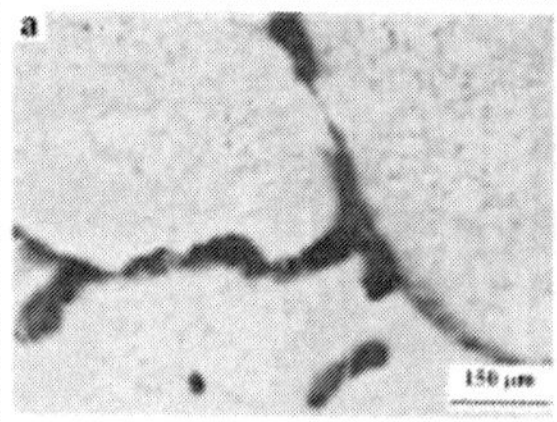 | 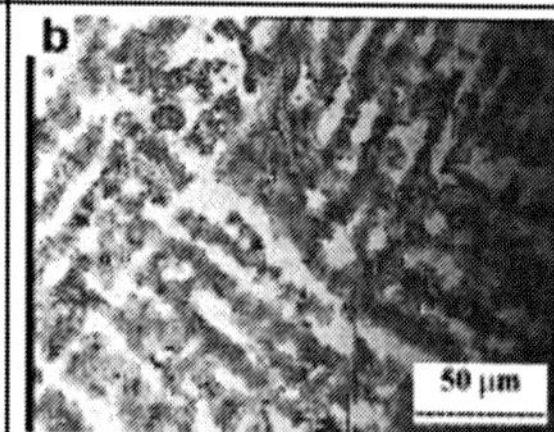 |
| HH2 | 950 | Air | 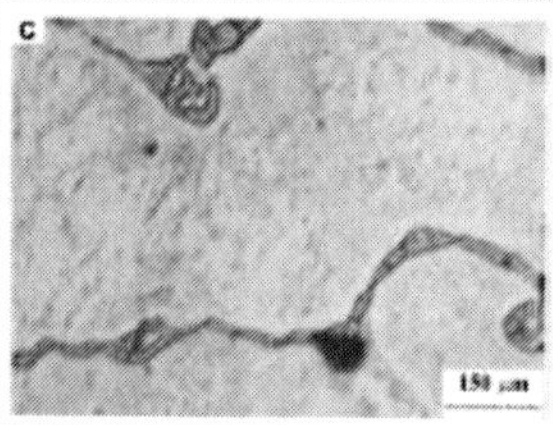 |  |
| HH8 | 1050 | Air | 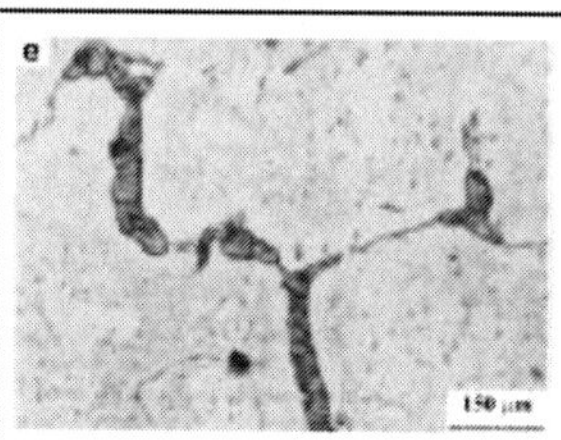 | 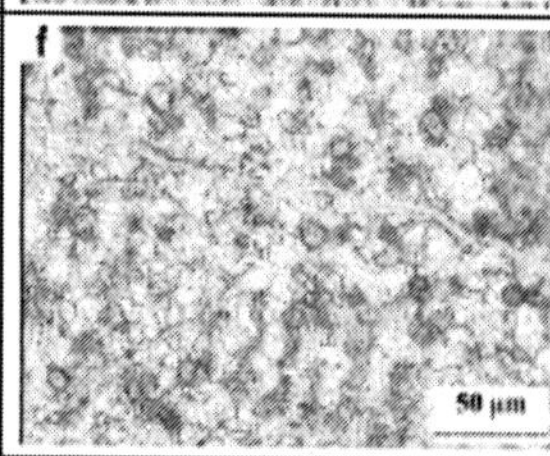 |
| HH11 | 1100 | Air | 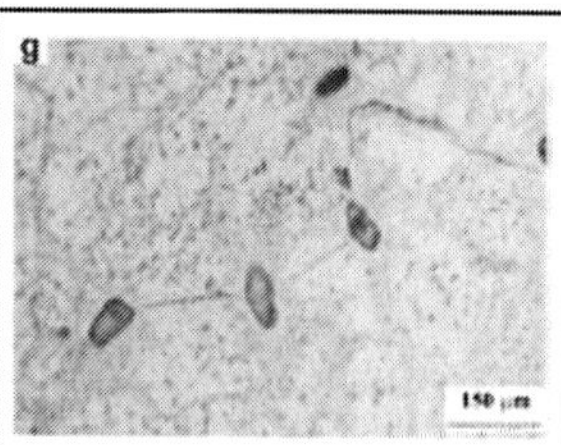 | 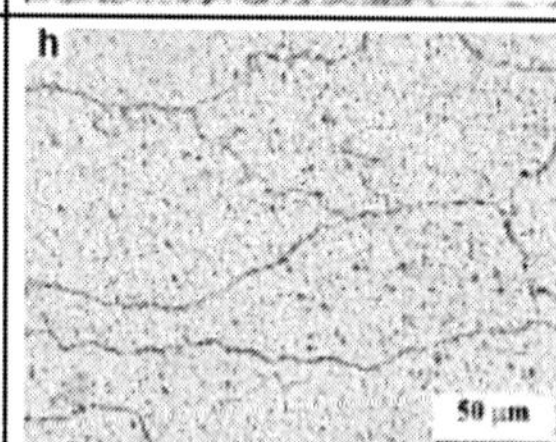 |
| HH14 | 1100 | Furnace | 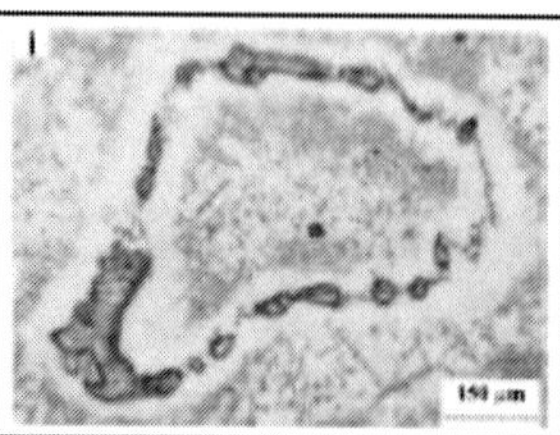 | 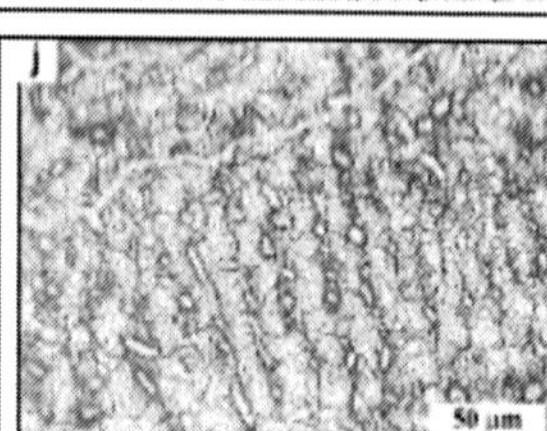 |

As can be seen in Fig. 11, the heat treatment at 1100°C followed by cooling in air led to changing the morphology of sigma phase from dendritic structure (Figs. 11(a) and (b)) to globular structure (Fig. 11(c)). The dendritic sigma phase indicates an unstable shape, and globular sigma phase exhibits a stable shape. Also, the dendritic structure is brittle, while, the globular structure is ductile. Therefore, it can be concluded that for completing the transformation of delta-ferrite to sigma phase, the temperature of homogenization heat treatment should be increased to 1100°C and the morphology of sigma phase be transformed to stable globular structure. This structure leads to remarkable increase in the impact energy and ductility of continuous annealing furnace roller. In contrast, the formation of sigma-phase with dendritic morphology results in decrease in ductility and failure of the roller.

Fig. 13 shows the fracture surfaces of the failed sample before homogenization (Fig. 13(a)), the sample after the homogenization heat treatment at 1100 °C for 2 h followed by cooling in furnace (Fig. 13(b)),and the sample after the homogenization heat treatment at 1100 °C for 2 h followed by cooling in air (Fig. 13(c)). The EDS analysis of points 1 and 2 in Figs. 13(a) and 13(b), respectively indicated the presence of sigma-phase in definite crystallographic planes. In fact, the fracture surface of failed sample (Fig. 13(a)) contained a random arrangement of flat surfaces and surfaces with steps which is characteristic of brittle fracture. The fracture surface of the sample after the homogenization heat treatment at 1100 °C for 2 h followed by cooling in furnace (Fig. 13(b)) was a mixture of small dimples and layered surfaces that indicated brittle and ductile fractures in this sample. The EDS analysis of point 3 in Fig. 13(b) showed the presence of chromium in the form of carbide particles. The carbide particles existed in the center of small dimples. According to EDS analysis of points 2 and 3, it can be said that due to slow cooling rate in the furnace, all the desirable changes that occur in homogenized structures are reversed and a new formation of the sigma-phase and precipitates in the sample results in a transition from ductile to almost brittle fracture. As can be seen in Fig. 13(c), formation of deep dimples is a main characteristic of ductile fracture. The EDS analysis of the point 4 was similar to the base metal. This figure indicated a considerable decrease in sigma phase and precipitates amounts after homogenization.

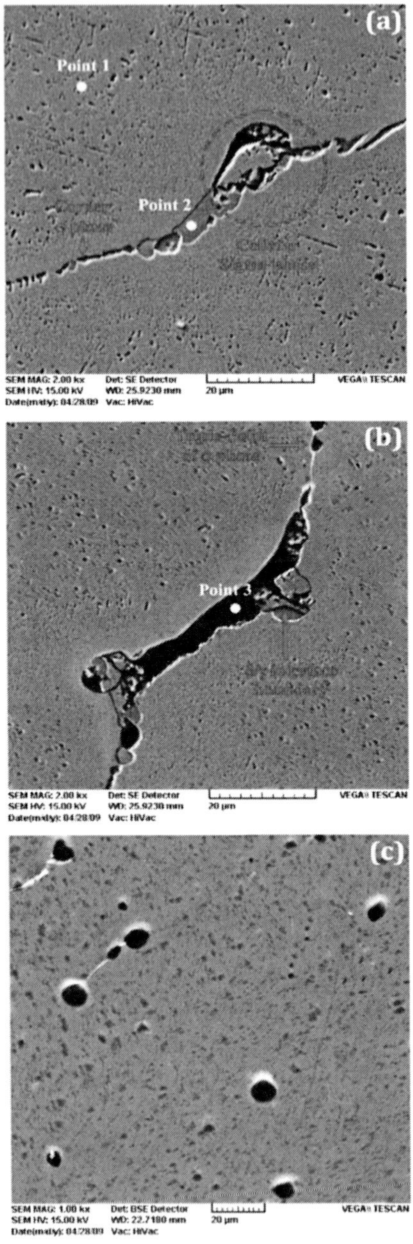


**Figure 10.** Morphology of sigma-phase. (a) and (b) after failure and before homogenization treatment, respectively and (c) after the homogenization heat treatment at 1100 °C for 2 h followed by air cooling.

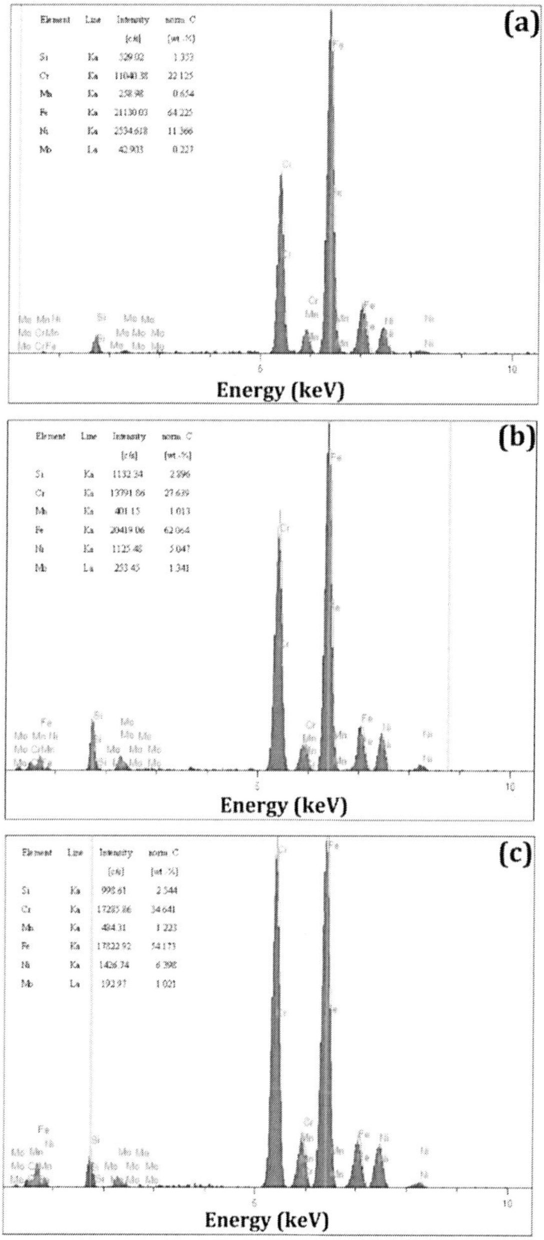


**Figure 11.** The EDS analysis of: (a) austenite-phase (point 1 in **Fig. 11(a)), (b)** ferrite-phase (point 2 in **Fig. 11(b)),** and (c) sigma-phase (point 3 in **Fig. 11(c))**.

**Table 2.** Elemental composition of **Fig. 12**.

| Point | Element (wt.%) | | |
|---|---|---|---|
| | Cr | Si | Mo |
| 1 (gamma) | 22.125 | 1.353 | 0.227 |
| 2 (delta) | 27.639 | 2.896 | 1.341 |
| 3 (sigma) | 34.641 | 2.544 | 1.021 |

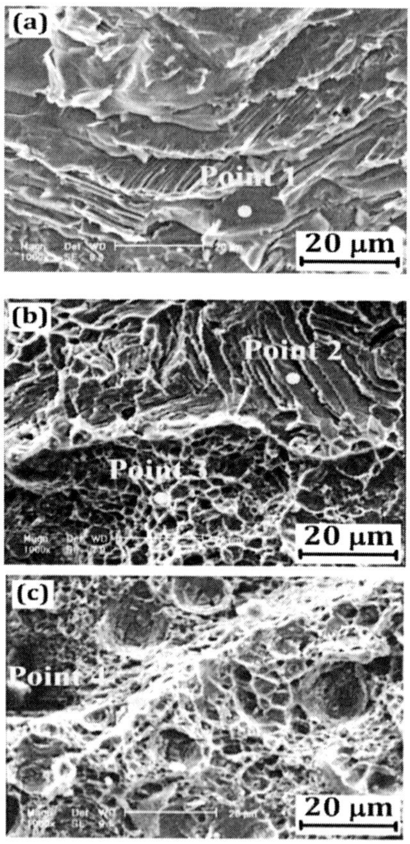


**Figure 12.** Fracture surfaces of: (a) the failed sample before homogenization, (b) the sample after the homogenization heat treatment at 1100 °C for 2 h followed by cooling in furnace, and (c) followed by cooling in air.

Therefore, before roller straightening, the homogenization heat treatment should be performed for dissolution of deleterious phases in austenite as the matrix phase and obtaining good toughness. In many cases, microstructural inhomogeneity can be eliminated by holding the sample at high temperatures for a sufficient time and cooling at an appropriate rate. Otherwise, the possibility of the formation of fine cracks during roller straightening is very high. These cracks can result in roller fracture during the straightening process and/or in the continuous-annealing furnace.

## 6. CONCLUSIONS

The sigma phase tends to precipitate in the regions of high chromium content, such as the chromium carbides in the grain. A relatively small quantity of the σ phase, when it is nearly continuous at a grain boundary, can lead to very early failure of high-temperature parts. It is important to understand that the sigma-phase cannot undergo any significant plastic deformation; instead, it fractures even at relatively low strain levels. This is true at elevated temperatures, but even more so at ambient temperature. Where toughness and ductility are an important part of the system design, the sigma-phase cannot be tolerated. The formation of sigma-phase with dendritic morphology results in decrease in ductility and failure of heat resistant steels. The microstructural inhomogeneity (sigma-phase and carbides) can be eliminated by holding the sample at high temperatures for a sufficient time and cooling at an appropriate rate. The homogenization heat treatment under appropriate condition can led to change in the morphology of sigma phase from dendritic structure to globular structure. This structure leads to remarkable increase in the impact energy and ductility of heat resistant steels.

## REFERENCES

1. A. S. M. Handbook, . Volume, Properties, Irons. Selection, Steels, Alloys. A. S. M. High-Performance, Materials. International, Park, Ohio, 2005

2. Lamb, S., Practical Handbook of Stainless Steels and Nickel Alloys, ASM International, Materials Park, Ohio, 1999.

3. ASM Handbook, Volume 15, Casting, ASM International, Materials Park,Ohio, 2005

4. S. Shi, J. C. Lippold, ". Microstructure, During. Evolution, Exposure. Service, Two. of, Heat. Cast-Resisting, Steels. Stainless-Hp-Nb, Modified, Materials. 20-32nb", Vol. Characterization, 810292008

5. Prager, M. and Svoboda, J., Cast High Alloy Mettalurgy, Steel Casting Metallurgy, Steel Founder's Society of America, Rocky River, OH, 1984.

6. W. Trietschke, G. Tammnann, Alloys. The, Iron. of, Chromium, Zh. Anorg. Chem., 551907402

7. E.C. Bain and W.E. Griffiths, An Introduction to the Iron-Chromium-Nickel Alloys, Trans.AIME, 751927166

8. P. Chevenard, Investigations. Experimental, Iron. of, Nickel, Alloys. Chromium, Trav, Mem., Bur. Int. Poids et Mesures, 17192790

9. F. Wever, W. Jellinghaus, Two. The-Component, Iron. System-Chromium, Kaiser. Mitt-Wilhelm, Inst, 1. Vol, 1931143

10. E. O. Hall, S. H. Algie, Sigma. The, Metall. Phase, Rev, 1. Vol, 196661

11. ASM Handbook, Volume 11, Failure Analysis and Prevention, ASM International, Materials Park, Ohio, 2002.

12. M. H. Bina, G. Dini, S. M. M. Vaghefi, A. Saatchi, K. Raeissi, M. Navabi, ". Application, Homogenization. of, Treatments. Heat, Improve. to, Furnace. Continuous-Annealing, Fractures". Roller, Failure. Engineering, Vol. Analysis, 517202009

13. G. Dini, M. H. Bina, S. M. M. Vaghefi, K. Raeissi, M. Safaei-Rad, M. Navabi, ". Failure, a. of, Furnace. Continuous-Annealing, at. Roller, Steel. Mobarakeh, Engineering. Company", Analysis. Failure, Vol, 78562008

14. A.J. Lena and W.E. Curry, The Effect of Cold Work and Recrystallization on the Formation of the Sigma Phase in Highly Stable Austenitic Stainless Steels, Trans.ASM, 471955193

15. ASM Handbook, Volume 19, Fatigue and Fracture, ASM International, Materials Park,Ohio, 2005

16. ASM Handbook, Volume 12, Fractography, ASM International, Materials Park,Ohio, 2005

17. Lin, D.Y., Liu, G.L., Chang, T.C. and Hsieh, H.C., "Microstructure Development in 24Cr-14Ni-2Mn Stainless Steel after Aging under Various Nitrogen/Air Ratios", Journal of Alloys and Compounds, Vol.1-21501542004

18. C. C. Tseng, Y. Shen, S. W. Thompson, M. C. Mataya, G. Krauss, ". Fracture, Formation. the, Sigma. of, M. Phase, C. , from. Austenite, in. Delta-Ferrite, A. I. S. I. 3. an, L. Stainless, Metallurgical. Steel", Transactions. A. Materials, Vol, 6114711581994

19. M. Martins, L. C. Casteletti, ". Sigma, Morphologies. Phase, Cast. in, Super. Aged, Stainless. Duplex, Materials. Steel", Vol. Characterization, 87927952009

20. C. C. Hsieh, D. Y. Lin, W. Wu, ". Precipitation, of. σ. Behavior, in. . Phase, Cr, Ni, Mn, . , Cr-0., Stainless. Si, Hot. Steels-Rolled, 8. at, °. C. with, Reduction. Various, Materials. Ratios", Science, A. Engineering, Vol, 1-21811892007

21. T. P. S. Gill, M. Vijayalkshmi, P. Rodriguez, K. A. Padmanabhan, ". On, Correlation. Microstructure-Property, Thermally. of, Type. 3. Aged, L. Stainless, Weld. Steel, Metallurgical. Metal", A. Transactions, Vol, 6111511241989

22. C. C. Hsieh, D. Y. Lin, T. C. Chang, ". Microstructural, During. Evolution, δ/σ/γ. the, Transformation. Phase, the. of, 3. Sus, L. Si, Steel. Stainless, Aging. after, Various. under, Atmospheric. Nitrogen, Materials. Ratios", Science, A. Engineering, Vol, 1-21281352008

23. G.N. Emanuel, Sigma Phase and Other Effects of Prolonged Heating at Elevated Temperatures on 25 Per Cent Chromium-20 Per Cent Nickel Steel, in Symposium on the Nature, Occurrence, and Effects of Sigma Phase, STP 110, American Society of Testing and Materials,195182

24. G. Aggen, et al.Versus. Microstructures, of. Properties, 29, Stainless. Ferritic, in. Steel, Con. 7. Mi, of. Optimization, Properties. Processing,

Performance. Service, Microstructural. Through, S. T. P. Control, 67, Society. American, Testing. for, Materials, 1979334

10

# DEVELOPMENT AND CHARACTERIZATION OF A METAL INJECTION MOLDING BIO SOURCED INCONEL 718 FEEDSTOCK BASED ON POLYHYDROXYALKANOATES

**Alexandre Royer *, Thierry Barrière and Jean-Claude Gelin**

Department of Applied Mechanics, FEMTO-ST Institute,
26 Rue de l'Epitaphe, 25000 Besançon, France

## ABSTRACT

The binder plays the most important role in the metal injection molding (MIM) process. It provides fluidity of the feedstock mixture and adhesion of the powder to keep the molded shape during injection molding. The binder must provide strength and cohesion for the molded part and must be easy to remove from the molded part. Moreover, it must be recyclable, environmentally friendly and economical. Also, the miscibility between polymers affects the homogeneity of the injected parts. The goal of this study is to develop a feedstock of superalloy Inconel 718 that is environmentally friendly. For these different binders, formulations based on polyethylene glycol (PEG), because of his water solubility property, and bio sourced polymers were studied. Polyhydroxyalkanoates (PHA) were investigated as a bio sourced polymer due to its miscibility with the PEG. The result is compared to a standard formulation using polypropylene (PP). The chemical and rheological behavior of the binder formulation during mixing, injection and debinding process were investigated. The feedstock was characterized in the same way as the binders and the interactions between the powder and the binders were also studied. The results show the well adapted formulation of polymer binder to produce a superalloy Inconel 718 feedstock.

## KEYWORDS

metal injection molding; Inconel 718; polyhydroxyalkanoates; polyethylene glycol

## 1. INTRODUCTION

Metal injection molding (MIM) is a process to produce on mass small parts of less than 50 g with complex geometry from varied materials such as stainless steel, superalloys, carbides or ceramics. The MIM process provides good dimensional accuracy (tolerances < 5 microns), a good surface finish (*R*a < 1.5 microns) and high mechanical strength similar to those of the powder material [1,2]. The MIM is a process that involves many industrial sectors such as automotive, aerospace, information technology or medical. This process is based on the injection of a fluid material composed of powder of the desired material for the final part, and of a binder of several polymers. The piece is then subjected to debinding to remove the binder and then the piece is sintered to obtain a dense part [2].

The binder plays the most important role in the MIM process. It has to be able to support an important load rate of powder, typically 60%, and to carry the powder in the mold. Enneti, Quinard and Tam [2,3,4] showed that the binder must give strength and cohesion to the molded part and be easily removed from the molded part, while being recyclable, environmentally friendly and economical. These authors also explained the importance of a low viscosity, good adhesion to the powder, no chemical reaction with the powder which can contaminate the powder or scissor the polymeric chains, and a low coefficient of thermal expansion for binders. The binders are generally composed of three components [2]. One provides the necessary fluidity (polyethylene glycol (PEG) or waxes are generally used). Another one serves to provide strength of the injected piece, for which polypropylene or polyethylene is commonly employed. The last one is a surfactant which prevents the possible aggregation of the powder particles. Stearic acid is often used.

The different polymers of the binders used in this study are choose due to their environmentally friendly behavior in the MIM process. First the PEG permits the realization a water debinding due to its solubility in water [2,5]. This is environmentally important because the solvent debinding is generally realized using hazardous chemical solvent. Moreover the water and the PEG can be separated and then the water recycled. The PEG is a biocompatible thermoplastic semi-crystalline homopolymer which his characteristics depend on the molecular mass. It contains terminal hydroxyl groups which provides water-solubility for molecular masses ranging from 400 to 40,000 $g \cdot mol^{-1}$ [5,6,7]. PEG is not soluble in hydrocarbons but is soluble in acetone and alcohol [5]. Accordingly, a PEG with a molecular weight of 20,000 $g \cdot mol^{-1}$ was chosen as it ensures solubility in water while keeping a sufficient binder viscosity.

The choice of the primary binder polymer is driven by its affinity with the PEG. Moreover, to reduce its environmental impact, bio-sourced polymer was choosen because the primary binder has to be removed by thermal debinding to make the shape of the piece as long as possible. In addition, the influence of the binder on both homogeneity and rheological properties of the feedstock has been established [8]. Taking into account all these aspects, polyhydroxyalcanoates (PHA) was choose because it is well known to be miscible with the PEG [5]. Moreover, PHA is one of the most common bio sourced polymers and grades are available for mold injection application [9]. The PHA is a polyester obtained directly from the bacterial metabolism. More than 250 species of bacteria synthesized PHAs granules size from 0.2 to 0.5 µm as carbon and energy storage materials under condition of limiting nutrients and in the presence of excess carbon source. PHA is a biodegradable and biocompatible polymer with good physical, mechanical and thermal properties [5,9].

In this study, a micro powder of Inconel 718 was investigated for an application in the MIM process. A micro powder was chosen because it is well adapted for µMIM experiments, including injection, debinding and sintering stages [3,10]. The use of Inconel for MIM process has been studied by Özgün [11] and a formulation was developed. Inconel superalloys are used in aviation, aerospace and nuclear power due to their high resistance to corrosion and oxidation but also for their excellent mechanical strength at high temperature [11]. Currently, the most popular superalloys used for the above applications are Inconel

718 and 625. Here, an Inconel 718 powder was chosen as the final material of the piece. The binders developed by Özgün [11] were composed of polypropylene (PP), carnauba wax, paraffin wax and stearic acid.

The process to remove the PEG using the $CO_2$ supercritical as solvent was studied. This process heated and pressurised the $CO_2$ below his critical point (304.1 K and 7.38 MPa). The supercritical $CO_2$ has the behavior of a gas: this facilitates the penetration of the $CO_2$ inside the samples, and of a liquid: this increases the solubility of the PEG in the $CO_2$. The debinding method using the $CO_2$ supercritical as solvent was first reported by Chartier *et al.* in 1995 [12] but the tests were performed on feedstocks made of different waxes and ceramics powders. Shimizu *et al.* [13] also used this method of debinding with feedstock made of paraffin wax. Federzoni *et al.* [14] were the first to use the $CO_2$ supercritical as solvent debinding method to remove the PEG. Federzoni *et al.* [14] showed that the PEG can be removed entirely by using 65 °C, 300 bar and 7 h as conditions of the process.

The goal of the present study is to develop an environmentally friendly binder formulation adapted to the use of a micro powder of Inconel 718 with the use of bio sourced polymers as binder and the $CO_2$ supercritical as solvent debinding method.

## 2. MATERIALS AND METHODS

Two grades of PHA are used, namely PHA and PHBV provided by NaturePlast and adapted to the injection process. The powder used is an Inconel 718 atomized by argon provide by Sandvik Osprey (Neath, UK). The powder was characterized by a granulometer laser LA-950 V2 provided by Horiba (Kyoto, Japan), the tests were realized in water.

The polymer blends are made in a twin screw mixer provided by Brabender (Duisburg, Germany) which has a volume capacity of 50 $cm^3$, a speed of 50 rpm and at a temperature of 180 °C. The temperature of the test is determined between the higher melting temperature of the polymer and the lowest temperature of degradation of the polymers. According to this requirement, the mixtures were made for all the different kinds of mixture at 180 °C. The higher melting temperature corresponds to the PHA at 175 °C and the lowest temperature of degradation at the SA at 180 °C. The speed of the mixture was determined according to previous work in our laboratory [15]. First, the primary binder and the PEG were introduced in the mixer and then the SA and the powder. The mixture is homogenized when the mixing torque is stable.

The rheological tests were performed on a capillary rheometer Rosand RH2000 provided by Malvern Instruments (Malvern, UK). The tests are realized at 180 °C, the temperature selected for injection, in order to characterize the viscoelastic properties of the binders. The feedstocks were characterized at shear rate corresponding to the usual injection shear rate, between 2000 $s^{-1}$ and 50,000 $s^{-1}$ in a range of shear rates matching those encountered in injection process. The injection was realized on an injection machine (Arburg, Loßburg, Germany) at 180 °C. Cylinders of 100 mm of length and 10 mm of diameter were injected.

The choice of the PEG was made due to his possibility to be removed by water. To study this method of debinding, a cylinder of the different mixtures were injected. Different methods of water debinding were tested. First the samples are put on a bath of heat water [16,17]. Two different temperatures were tested, 50 °C and 60 °C, and the test were made during 24 h and 48 h because these conditions correspond to removal of the entire PEG from the feedstock made with the PP. Another test was made using a magnetic stirrer in the water bath [16,18]. To compare the result with the previous test, the same temperature and time were studied. Finally, an ultrasonic bath was used [18]. The ultrasonic bath use a 35 KHz and 50 W ultrasonic waves. Images of the samples before and after debinding were also made to characterize the possible defaults which can appear during the different process. To determine the weight loss of PEG the samples are weighed before and after each test. This is an approximation but only the PEG is logically soluble in water so the results are close to reality. To confirm the results, some TGA on the samples have to be made.

According to the work of Federzoni *et al.* [14] a preliminary test was made at 70 °C, 300 bar during 4 h. However, a previous test on the samples injected with the feedstock with PP shows only 40% of weight loss of PEG. Another test at 150 °C and 400 bars during 4 h was made and the results were better. These conditions were choosen because, according to the data from the National Institute of Standards and Technology [19], the density and the viscosity of the supercritical $CO_2$ are similar at 70 °C, 300 bar (257.34 kg/m$^3$, 26.004 μPa·s) and 150 °C, 400 bar (259.50 kg/m$^3$, 29.180 μPa·s). So the coefficient of diffusivity and solubility of the PEG in the $CO_2$ does not change because of the properties of the $CO_2$ but due to the increase of the temperature and the pressure that enhance these coefficients. The supercritical debinding was realized in a SFE2.2 supercritical fluid reactor provided by Separex (Champigneulles, France).

# 3. RESULTS

## 3.1. Powder Characterization

The Inconel 718 powder was characterized by laser granulometer to determine the particle size and distribution. The Table 1 gives the $D_{10}$, $D_{50}$ and $D_{90}$ of the powder and the Figure 1 gives the distribution and the cumulative distribution. The size of the powder is in accordance with the requirements of the MIM process with a size smaller that 20 µm. The distribution is relatively thin with a span of 1.35 µm. This shows that the powder chosen is adapted to our process.

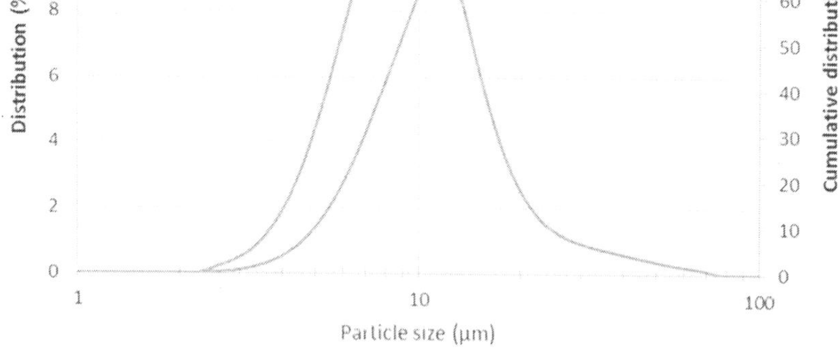


**Figure 1.** Size distribution (**a**) and cumulative volume (**b**) of the Inconel 718 powder.

**Table 1.** Characteristics of the Inconel 718 powder.

| Particles Size Distribution | $D_{10}$ | $D_{50}$ | $D_{90}$ |
|---|---|---|---|
| Size | 4.95 μm | 8.7 μm | 16.7 μm |

## 3.2. Mixing

To characterize the different mixtures (Table 2), the critical volume fraction of powder was determined for each formulation [3,20]. The results are given Table 3 and Figure 2. To determine the critical volume fraction of powder, the incremental method was used. This method consists of adding the powder progressively until the mixing torque cannot stabilize. The maximum volume fraction of powder corresponds to the last mixture that is stabilized. The results for the mixture show a lower critical volume fraction of powder for the In718/PEG/PP/SA mixture. The volume fraction of powder determines the value of the shrinkage during the sintering and also the uniformity of the shrinkage. A high value of volume fraction reduces the value of the shrinkage because the distance between the particles of powder become lower and less porosity remains after debinding. The mixtures based on PHBV and PHA are interesting at this point. The critical volume fraction was determined to characterize and compare the different mixtures but the other analyses were made on feedstock with 60% of volume fraction. This fraction was chosen because it is a usual volume fraction in industry.

## 3.3. Rheology

The feedstocks were characterized by capillary rheometer to determine their ability to be injected. The results are shownFigure 3. The results shows pseudo plastic behavior of all the mixtures and the results are in accordance with the mixing results. The pseudo plastic behavior of the feedstocks can be confirmed using the flow behavior index of the Ostwald's law (1). The results are available in Table 4 and the value of the flow behavior index of all the feedstocks corresponds to a pseudo plastic behavior (<1). Moreover the feedstock with PHA show low shear viscosity. This low shear viscosity can be easy to inject and makes it easier to fill the thin part of the component. The two other feedstocks have similar shear viscosity.

$$\eta = K\dot{\gamma}^{n-1} \tag{1}$$

Ostwald's law of fluid where n is the flow behavior index, η is the shear viscosity (Pa·s), *K* is the flow consistency index and γ˙ is the shear rate ($s^{-1}$).

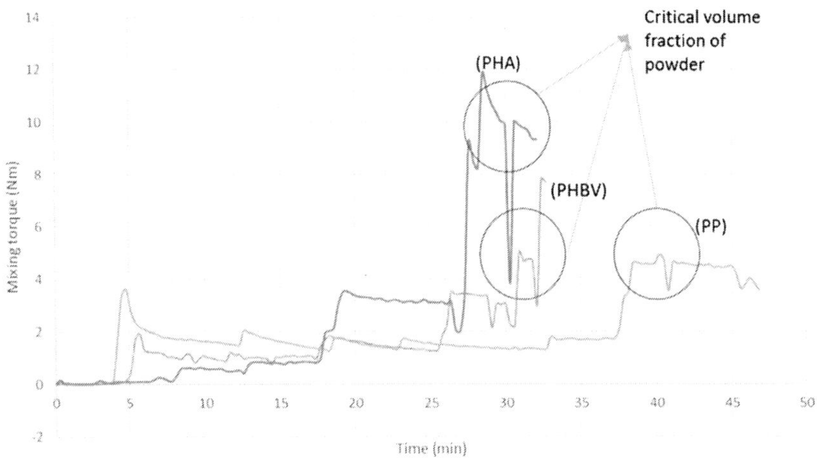


**Figure 2.** Mixing torque *versus* time of the feedstocks with polyhydroxyalkanoates (PHA), Poly(3-hydroxybutyrate-co-3-hydroxyvalerate) (PHBV) and polypropylene (PP) during the incremental method.

**Table 2.** Composition of the formulations.

| Volume Fraction (%) | Materials of Feedstock 1 | Materials of Feedstock 2 | Materials of Feedstock 3 |
|---|---|---|---|
| 60 | Inconel 718 | Inconel 718 | Inconel 718 |
| 22 | PEG20K | PEG20K | PEG20K |
| 16 | PP | PHA | PHBV |
| 2 | Stearic Acid | Stearic Acid | Stearic Acid |

**Table 3.** Critical volume fraction of powder and maximum volume fraction of powder of the different feedstock.

| Formulation | Critical Volume Fraction of Powder (%) | Maximum Volume Fraction of Powder (%) |
|---|---|---|
| In718/PEG/PHA/SA | 92 | 90 |
| In718/PEG/PHBV/SA | 87 | 85 |
| In718/PEG/PP/SA | 80 | 78 |

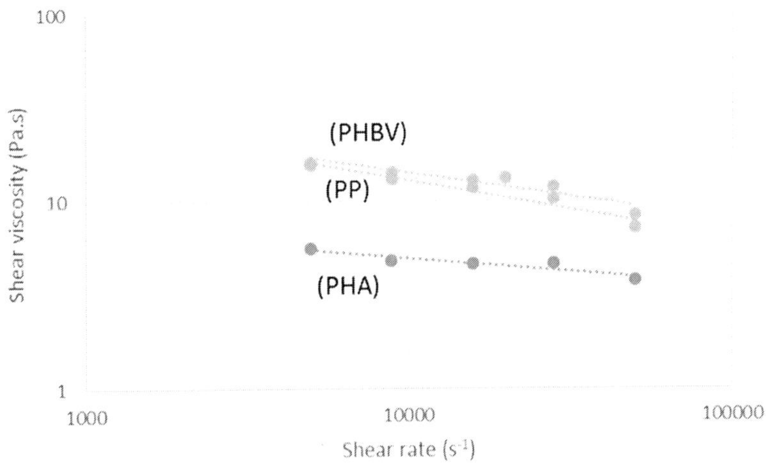


**Figure 3.** Shear viscosity *versus* shear rate of the feedstock with PP, PHBV and PHA and 60% of volume fraction of powder.

**Table 4.** Value of the flow behavior index of the different feedstock.

| Formulation | PP | PHA | PHBV |
|---|---|---|---|
| Value of the flow behavior index $n$ | 0.686 | 0.855 | 0.743 |

## 3.4. Water Debinding

The different results of the debinding test in water are available in Figure 4 and Figure 5. In water, the weight loss of PEG of the feedstocks with PP and PHBV are similar and after 48 h at 60 °C the PEG is totally removed. The PHA lost only 40% of PEG after this process. This lower weight loss of PEG can be explained by the affinity, or even miscibility, between the PEG and the PHA. This can induce a smaller inclusion of PEG in the PHA compared with PHBV or PP. The PEG is so much more difficult to access for the water and the time of debinding increase. The images of the cylinder with PHA may confirm this hypothesis because some cracks appear on the surface and may be due to the difficulty for the water to access to the PEG. Moreover, the use of 60 °C as temperature of the water reveals better results for the PHBV and PP mixture, contrary to the PHA feedstock that lost more PEG at 50 °C.

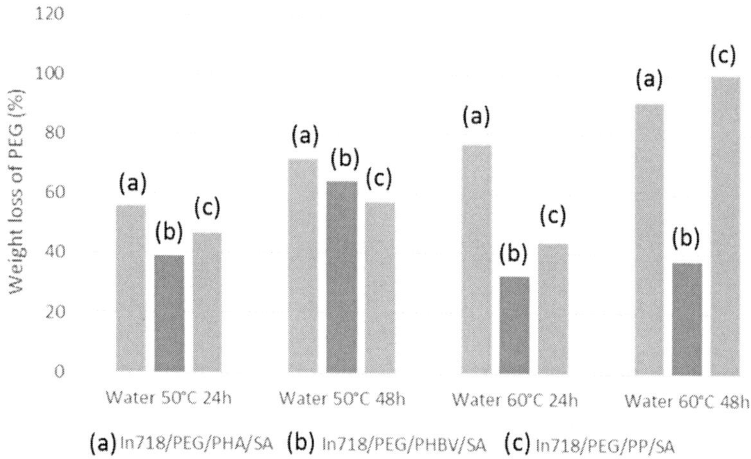


**Figure 4.** Weight loss of polyethylene glycol (PEG) for the different feedstocks after debinding in water at 50 °C and 60 °C during 24 h and 48 h.

The water debinding test with magnetic stirrer shows better results for each formulation. After 48 h at 60 °C all the formulations have lost 100% of PEG. Moreover, after 24 h at 60 °C the components have lost more PEG than after 24 h in water at 60 °C without stirring. This is probably due to the replacement of the water around the samples. This refresh water has a better possibility to remove the PEG. The results for water at 50 °C are similar to the previous test in water. However, the images show more cracks for the samples with PHA and PHBV. This is probably because more tension appears on the samples due to the mobility of the water.

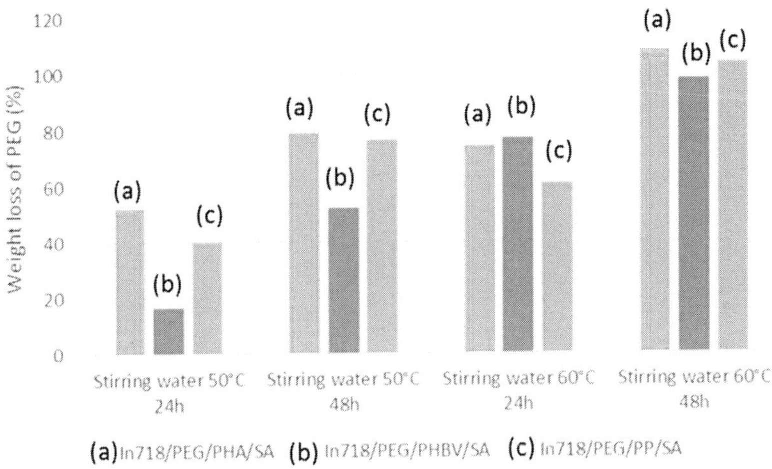


**Figure 5.** Weight loss of PEG for the different feedstocks after debinding in stirring water at 50 °C and 60 °C during 24 h and 48 h.

The debinding by ultrasonic water was studied and the results show between 150% and 350 % of weight loss of PEG. This is due to the loss of some powder during the process. Moreover, as shown in the images in Figure 6 the samples made with the feedstock with PHA or PHBV isare seriously damaged. The sample of the feedstock with PP does not show any default or crack. The tests were made in a bath fill with water at 60 °C.

## 3.5. Supercritical Debinding

The results of the different tests are shown in Figure 7 and images of the samples after the tests are available in Figure 8. The results show better weight loss for the feedstock with PP. The feedstock with PHA was totally destroyed during this process. This is probably due to the more fragile backbone of the PHA induced by the littlest inclusion of PEG. This fact decreases the strength of the sample, which, with the high temperature, cannot maintain the shape of the component. Regarding the others feedstocks, this process is softer than the debinding by water because the samples show no cracks or deformations.

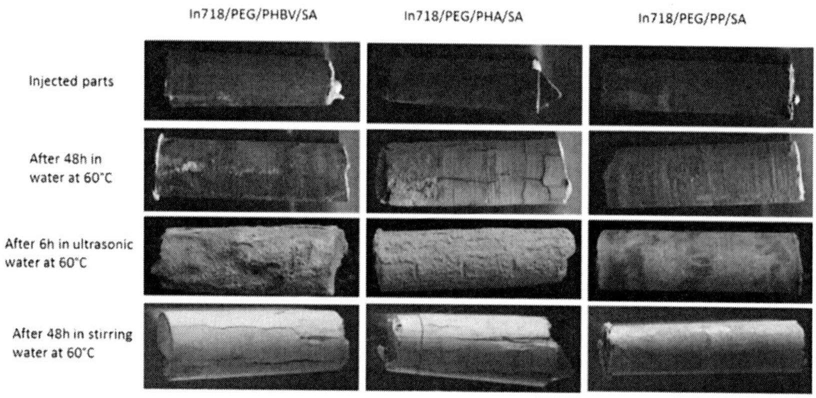


**Figure 6.** Images of the different samples after injection, after 48 h of debinding in water at 60 °C, after 6 h of debinding in ultrasonci water at 60 °C, and after 48 h of debinding in stirring water at 60 °C.

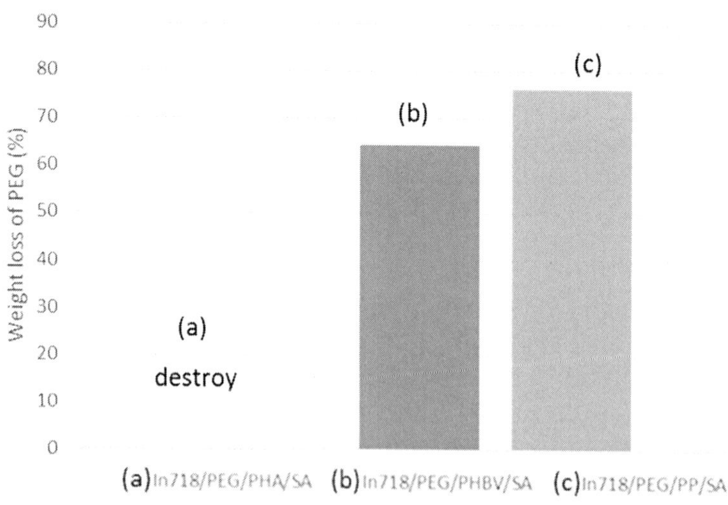


**Figure 7.** Weight loss of PEG for the different feedstocks after supercritical $CO_2$debinding process at 150 °C and 400 bar during 4 h.

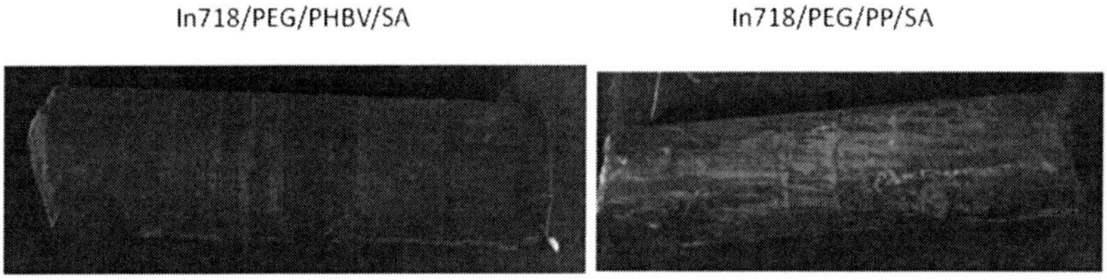


**Figure 8.** Images of the different samples after 4 h at 150 °C and 400 bar of supercritical $CO_2$ debinding.

# 4. DISCUSSION

The results show the PP can be substituted by the PHBV in a MIM feedstock. The results obtained with the feedstock with PHA show that the properties of this feedstock are not compatible with the MIM process. The PHA could be a primary binder but maybe the ratio with the PEG should be changed for a higher rate of PHA. The change of the ratio could more deeply modify the properties of the feedstock due to the possible modification of the parameters of miscibility between the PEG and the PHA. This could avoid the cracks, reduce the time of debinding and increase the strength of the component. This could be interesting because the viscosity of the feedstock with PHA is relatively low and an increase of the rate of PHA should not make the mixture too viscous to inject. Some analyzes like TGA or DSC will be done to analyze the possible miscibility between the PHA and the PEG and so optimize the ratio of the mixture. Finally, maybe the PHA should be mixed with another polymer which is more adapted.

The conditions of the $CO_2$ supercritical debinding process have to be optimized to successfully entirely remove the PEG in a short time. Maybe the increase of the time is not the best solution; the influence of the temperature and of the pressure could be studied more. A lower temperature can probably reduce the defaults because the polymers will be more resistant to the mechanical stress. According to our composition of feedstock, a test around 100 °C should be more compatible. This temperature is higher than the melting temperature of the PEG and sufficiently lower than the melting temperature of the PP. Moreover, a higher pressure between 400 bars and 600 bars should be tested because our equipment is limited to 600 bars. This would determine the influence of pressure on the debinding process. The influence of the different polymers and of the different debinding processes on the final densified part is an important study. This will be studied in future works because the modification of the surface of the samples by the polymers or the process can induce some problems during the thermal debinding and also during the sintering.

## 5. CONCLUSIONS

The results show that the PHBV can be a solution to substitute the PP and realize a green feedstock. Indeed, the behavior of the feedstock made with PHBV shows similar rheological properties, better maximal volume fraction of powder and no more defaults after debinding process. The time of debinding is longer but the use of the $CO_2$ supercritical debinding process reduces this time which becomes much shorter than for the water debinding. Moreover, this debinding process is smoother for the components during debinding. The feedstock made with PHA is interesting due to his low viscosity but too much default appears during debinding and the time to remove the PEG is too long. The tests of debinding have to be confirmed by TGA and the influence of the different process and the different polymers on the final densified piece will be studied.

## ACKNOWLEDGMENTS

The authors wish to thank the FUI ProPIM project for the financial support.

## AUTHOR CONTRIBUTIONS

Alexandre Royer designed and performed the experiments. Alexandre Royer analyzed the data and wrote the paper. Thierry Barrière and Jean-Claude Gelin supervised the study and corrected the paper.

## REFERENCES


1. German, R.M. *Powder Injection Moulding—Design and Applications*; Innovative Material Solution, Inc.: State College, PA, USA, 2003.

2. Enneti, R.K.; Onbattuvelli, V.P.; Atre, S.V. Powder binder formulation and compound manufacture in metal injection molding (MIM). In *Handbook of Metal Injection Molding*; Heaney, D.F., Ed.; Woodhead Publishing Series in Metals and Surface Engineering; Woodhead Publishing: Cambridge, UK, 2012; pp. 64–92.

3. Quinard, C.; Song, J.; Barriere, T.; Gelin, J.C. Elaboration of PIM feedstocks with 316L fine stainless steel powders for the processing of micro-components. *Powder Technol.* **2011**, *208*, 383–389.

4. Tam, K.C.; Yap, S.P.; Foong, M.L.; Loh, N.H. Metal injection molding: Effects of the vinyl acetate content on binder behavior. *J. Mater. Process. Technol.* **1997**, *67*, 120–125.

5. Thomas, S.; Grohens, Y.; Jyotishkumar, P. *Characterization of Polymer Blends: Miscibility, Morphology and Interfaces*; John Wiley & Sons: New York, NY, USA, 2014.

6. Sheth, M.; Kumar, R.A.; Davé, V.; Gross, R.A.; McCarthy, S.P. Biodegradable polymer blends of poly(lactic acid) and poly(ethylene glycol). *J. Appl. Polym. Sci.* **1997**, *66*, 1495–1505.

7. Castillo, R.V.; Müller, A.J. Crystallization and morphology of biodegradable or biostable single and double crystalline block copolymers. *Prog. Polym. Sci.* **2009**, *34*, 516–560.

8. Thavanayagam, G.; Pickering, K.L.; Swan, J.E.; Cao, P. Analysis of rheological behaviour of titanium feedstocks formulated with a water-soluble binder system for powder injection moulding. *Powder Technol.* **2015**, *269*, 227–232.

9. Jiang, L.; Zhang, J. Biodegradable Polymers and Polymer Blends. In *Handbook of Biopolymers and Biodegradable Plastics: Properties, Processing and Applications*; Ebnesajjad, S., Ed.; William Andrew Publishing: Boston, MA, USA, 2013; pp. 109–128.

10. Quinard, C.; Barriere, T.; Gelin, J.C. Development and property identification of 316L stainless steel feedstock for PIM and μPIM. *Powder Technol.* **2009**, *190*, 123–128.

11. Özgün, Ö.; Gülsoy, H.Ö.; Yılmaz, R.; Fındık, F. Microstructural and mechanical characterization of injection molded 718 superalloy powders. *J. Alloy. Compd.* **2013**, *576*, 140–153.

12. Chartier, T.; Ferrato, M.; Baumard, J.-F. Influence of the debinding method on the mechanical properties of plastic formed ceramics. *J. Eur. Ceram. Soc.* **1995**, *15*, 899–903.

13. Shimizu, T. Solvent debinding of injection moulded parts with supercritical carbon dioxide. *Met. Powder Rep.* **1997**, *52*, 41.

14. Federzoni, L.; Revirand, P.; Lumia, D.; Rossignol, G. Evaluation of the supercritical CO2 as method of debinding for PIM parts. In Proceedings of the Internationnal Congress & Exibition, Euro PM, Copenhagen, Denmark, 12–14 October 2009.

15. Djoudi, H.; Gelin, J.-C.; Barrière, T. Experiments and FE simulation for twin screw mixing of nanocomposite of polypropylene/multi-walled carbon nanotubes. *Compos. Sci. Technol.* **2015**, *107*, 169–176.

16. Chen, G.; Cao, P.; Wen, G.; Edmonds, N. Debinding behaviour of a water soluble PEG/PMMA binder for Ti metal injection moulding. *Mater. Chem. Phys.* **2013**, *139*, 557–565.

17. Hidalgo, J.; Jiménez-Morales, A.; Barriere, T.; Gelin, J.C.; Torralba, J.M. Water soluble Invar 36 feedstock development for µPIM. *J. Mater. Process. Technol.* **2014**, *214*, 436–444.

18. Shbeh, M.M.; Goodall, R. Design of water debinding and dissolution stages of metal injection moulded porous Ti foam production. *Mater. Des.* **2015**, *87*, 295–302.

19. US Department of Commerce. National Institute of Standards and Technology. Available online: http://www.nist.gov/(accessed on 25 March 2016).

20. Hidalgo, J.; Abajo, C.; Jiménez-Morales, A.; Torralba, J.M. Effect of a binder system on the low-pressure powder injection moulding of water-soluble zircon feedstocks. *J. Eur. Ceram. Soc.* **2013**, *33*, 3185–3194

77

# MOULD FILLING ABILITY CHARACTERISATION OF CAST ALUMINIUM ALLOYS USING DESIGN OF EXPERIMENTS

**Samavedam Santhi,[1] S. B. Sakri,[2] Dharwada Hanumantha Rao,[3] and Srinivasan Sundarrajan[4]**

[1]Department of Metallurgical and Materials Engineering, MGIT, Hyderabad 500075, India

[2]Special Fabrication Division, DRDL, Hyderabad 500058, India

[3]Mechanical Engineering Department, M.V.S.R. Engineering College, Hyderabad 501510, India

[4]National Institute of Technology, Trichy 620015, India

## ABSTRACT

Mould filling ability characteristic has been influencing the liquid metal flow, heat transfer, and solidification there by affecting the quality of aerospace castings. Mould filling is a critical parameter in the production of sound and quality castings, especially in the case of complex-shaped castings where section thickness is varying considerably. The mould filling ability of aluminium alloys LM6 and LM25 has been studied in the present investigation. Experimentation has been carried out using orthogonal array experimental layout. The process parameters studied in the present investigation are alloy composition, sand fineness, mould coating, and pouring temperature. The mould filling ability of selected aluminium alloy has been studied using pin test piece with cylindrical cores. The results from experimentation are analyzed to find the influence of the process parameters on mould filling ability. Based on the above, LM6 alloy has been found to have better mould filling ability characteristics and the analysis of variance has also revealed the same optimum factor combination.

# 1. INTRODUCTION

Metal casting is unique among metal forming processes for a variety of reasons [1]. Aluminum castings have played an integral role in the growth of the aluminum industry since its inception in the late 19th century. Popular aluminum casting alloys contain, in addition to strengthening elements, sufficient amounts of eutectic-forming elements (usually silicon) in order to have adequate fluidity to make the metal flow through the cavities. The aluminum castings and applications rapidly expanded to address the requirements of a wide range of engineering specifications and continuing market requirement to reduce vehicle weight and to increase fuel efficiency. Complicated physics involved in the casting process is fluid mechanics with phase change, shrinkage, porosity, macrosegregation in alloys, heat transfer between metal and mould, and thermal stress in the solidifying ingot [2]. Thin-wall aluminium alloy casting techniques can be used to produce clean castings of lower weight and with superior mechanical and physical properties compared to those produced by more complex traditional methods and are ideal for automotive applications [3]. The filling of the mould is an essential part of the entire casting cycle. Mould-filling ability is the bottleneck of manufacturing thin-walled castings [4].

The solidification of the melt during the mould filling continuously changes the properties of liquid and solid phases. The filling conditions play a significant role on the casting quality. A good mould-filling metal has the ability to fill out the mould pattern and reproduce fine details of the mould [5, 6]. Many defects in a casting generally originate at the filling stage [6].

The gating design is driven by the ideal mould filling time, which depends on the cast metal, casting weight, and minimum wall thickness [7]. Fast filling leads to turbulence-related defects (such as mould erosion, air aspiration, and inclusions). On the other hand, slow filling may cause defects related to premature solidification (such as cold shuts and misruns).

The integrity of a casting will be degraded substantially if high levels of free surface turbulence are experienced by the liquid metal during mould filling [8, 9]. This is due to the entrapment and breakup of surface oxide film formed on the surface of liquid metal. This oxide film becomes trapped in the casting and behaves like a crack. This generally appears when large flat castings filled horizontally [10].

Mould filling is strongly dependent on the surface tension and viscosity of the metal. It is affected by the mould characteristic, melt composition, and melt cleanliness. The mould filling process that closely related to molten metal flow, heat transfer, and mass transfer process is a flow process at changeable temperature accompanied with heat loss and solidification [4]. Factors affecting mould filling generally are alloy composition, mould, sand fineness, wall thickness, mould coat [7], gating system [7, 11], pouring issue (pouring temperature), and filters [12].

Taguchi's methods belong to the class of approaches that attempt to ensure quality through design. The design of experiments using Taguchi's approach can economically satisfy the needs of problem solving and product/process design optimization projects in the manufacturing industry. Multiple process variables can be studied at the same time with Taguchi's orthogonal array-efficient designs, instead of in a hit-and-miss approach, providing very reproducible results. Due to the statistical balance of the designs, more potential combinations of process variables (at different levels) can be evaluated for the best overall combination, in small number of experiments.

Orthogonal array technique allowed many factors to be analysed with limited number of experiments. Orthogonal array of design of experiments makes the design to be balanced so that factor levels are weighted equally [5]. Because of this, each factor can be evaluated independently of all the other factors, so the effect of one factor does not influence the estimation of another factor. MINITAB software is used for experimental design [13, 14]. Orthogonal arrays provide a set of well-balanced (minimum) experiments.

## 2. EXPERIMENTAL PLAN

Designing of experiments is an efficient approach for improving a process. Changing more than one factor at a time to quickly obtain meaningful results and draw conclusions about how factors interact is the primary objective of the design of experiments. A full scale of experimentation requires 16 experiments which have been reduced to 8 experiments using orthogonal array design of experiments.

For studying mould filling ability, the influencing parameters selected are alloy composition, sand fineness, mould coat, and pouring temperature.

Alloy composition affects the fluidity, filling characteristics, mechanical properties, and structure of the cast product. To study the influence of alloy composition, two alloys are taken for the present study; these are LM6 and LM25. Table 1 gives the chemical composition of the two alloys.

Another processing parameter is fineness of the sand which shows significant difference in the casting characteristics like surface finish and strength. AFS sand fineness numbers 25 and 40 are considered.

The third processing parameter is mould coat [5]. Generally mould coat provides smooth surface. Pouring temperature [12] influences fluidity, porosity, strength, and structure of the casting. Hence, pouring temperature with 20°C of super heat is considered as one of the process parameters for the present study.

Table 2 shows the details of the factors and their levels for the present study. Orthogonal array L8 (2**4) is used with four factors and 2 levels (for 8 runs) for experimental investigation as shown in Table 3.

## 3. EXPERIMENTAL DETAILS

The mould filling ability of these alloys has been studied using pin test piece with cylindrical cores designed by Engler and Ellerbrok as shown in Figure 1. 50 mm diameter cylindrical cores are considered for the present study. The test casting consists of two fins of metal at both sides. The inverse of the diameter of curvature of the edge tip of the fin gives the value of the mould filling ability. The mould filling ability pin test piece is based on the determination of the diameter of meniscus of liquid metal just at the time of solidification in the test casting. It is possible to quantitatively determine the values of mould filling ability at various pressure heads using the pin test piece. The mould filling ability as the percentage length of the spiral which has successfully reproduced the contour to the total length gives only empirical values.

**Table 1.** Chemical composition (% wt).

| Alloy | Si | Fe | Cu | Mn | Mg | Ni | Zn | Pb | Sn | Ti | Each | Total | Al |
|---|---|---|---|---|---|---|---|---|---|---|---|---|---|
| LM6 | 10.5–13.5 | 0.65 | 0.15 | 0.55 | 0.1 | 0.1 | 0.15 | 0.1 | — | 0.2 | 0.05 | 0.15 | Rem |
| LM25 | 6.5-7.5 | 0.55 | 0.2 | 0.35 | 0.2–0.65 | 0.15 | 0.15 | 0.15 | 0.05 | 0.05–0.25 | 0.05 | 0.15 | Rem |

**Table 2.** Factors and their levels.

| | Factor 1 Alloy | Factor 2 Sand fineness (AFS No.) | Factor 3 Mould coat | Factor 4 Pouring temperature (°C) |
|---|---|---|---|---|
| Level 1 | LM25 | 25 | Graphite | $T$ |
| Level 2 | LM6 | 40 | No coating | $T + 20°C$ |

**Table 3.** L8 Orthogonal array.

| Exp. run order | Alloy | Sand fineness (AFS no.) | Mould coat | Pouring temperature (°C) |
|---|---|---|---|---|
| 1 | LM25 | 25 | Graphite (GC) | $T$ |
| 2 | LM25 | 25 | No coating (NC) | $T + 20$ |
| 3 | LM25 | 40 | Graphite (GC) | $T$ |
| 4 | LM25 | 40 | No coating (NC) | $T + 20$ |
| 5 | LM6 | 25 | No coating (NC) | $T$ |
| 6 | LM6 | 25 | Graphite (GC) | $T + 20$ |
| 7 | LM6 | 40 | Graphite (GC) | $T + 20$ |
| 8 | LM6 | 40 | No coating (NC) | $T$ |

## 3.1. Preparation of Green Sand Mould

Testing arrangement for the mould filling ability is shown in Figure 2. Tooling for experiments and assembled mould for this experiment are given in Figure 2. The mould filling pattern and cylindrical cores are provided in the cope box and it is shown in Figure 2(a). The cope box, drag, and cylindrical cores are given in Figure 2(b). Top view of the assembled mould for mould filling ability experiments is shown in Figure 2(c).

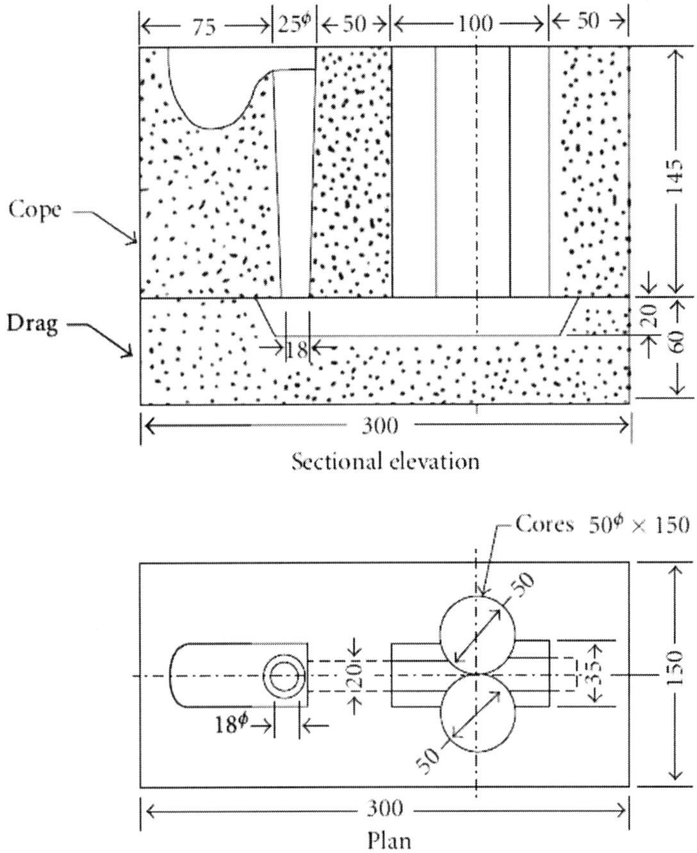


**Figure 1.** Schematic diagram of mould used for experimentation.

The moulds are provided with dowel pins for perfect matching of cope and drag. Moulds are prepared with slight ramming. The patterns have been stripped after 3 hours.

Moulds are prepared using green sand process. The composition consists of Bentonite 5-6% of sand weight and water is 5–8% of sand weight. In case of mould coatings the graphite paint is sprayed on to the mould and mould is dried immediately by lightening a flame on the painted surface.

(a) Mould filling pattern and cylindrical cores (b) Cope box, drag, and cylindrical cores (c) Assembled mould for mould filling ability

**Figure 2.** Tooling for mould filling experiments and assembled mould for this experiment.

## 3.2. Alloys

Alloys used in the present investigation are LM6 (Al-12Si) and LM25 (Al-7Si-0.4 Mg). Each alloy melting done with their respective same batch (belong to same heat) for all the 8 castings.

## 3.3. Melting and Pouring

The alloys are melted in an electric resistance furnace of capacity 20 Kg provided with mild steel crucible. Temperature is measured with the help of a thermocouple. The furnace is put off and the crucible is lifted and put in a tilting device. The metal is tapped into a smaller crucible for pouring into the mould.

## 3.4. Process Variables Studied

The variables studied are alloy composition, sand fineness, mould coat, and pouring temperature.

# 4. ANALYSIS OF RESULTS

Metal filling is an important phase of the casting process. The ability to fill and the contours play a significant role in reproducing the intricate details of the casting. The mould filling ability depends on the metallostatic balance between the metal pressure head and the surface tension of the mould. H is the height of liquid metal column above the filling point. A higher metallostatic pressure gives higher velocity of molten metal and thereby higher fluidity. The casting fluidity and filling are driven by metallostatic pressure head.

The test casting obtained for experimental run order 7 is shown in Figure 3.

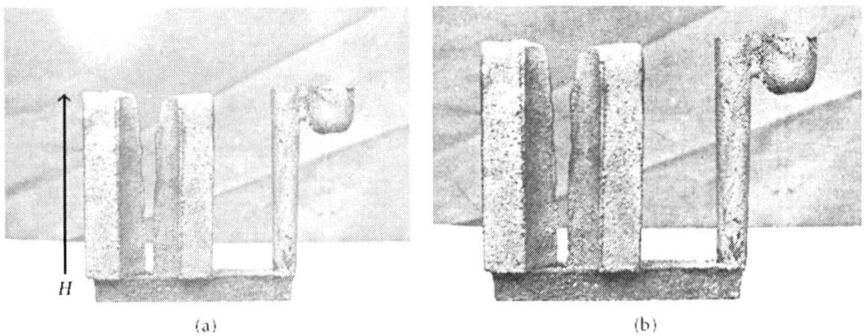


**Figure 3.** Mould filling ability test casting.

Using this test piece the mould filling ability value can be determined at various pressure heads (Hs) as shown in Figure 3. Pressure head is the height of liquid metal column above the filling point. The casting consists of two fins of metal at both sides.

The hot metal poured into the sprue fills the curved cavity between the two cylindrical cores having a line contact at the center. But it solidifies before it fills up the complete casting. The inverse of the diameter of curvature of the edge tip of the fin gives the value of the mould filling ability. The diameter at the tip of the fin gives the meniscus diameter of the liquid metal at the time of solidification as represented in Figure 4 for one particular pressure head.

It is difficult to measure the diameter of the tip of the edge and hence an indirect way of calculation has been used. As per Figure 5,

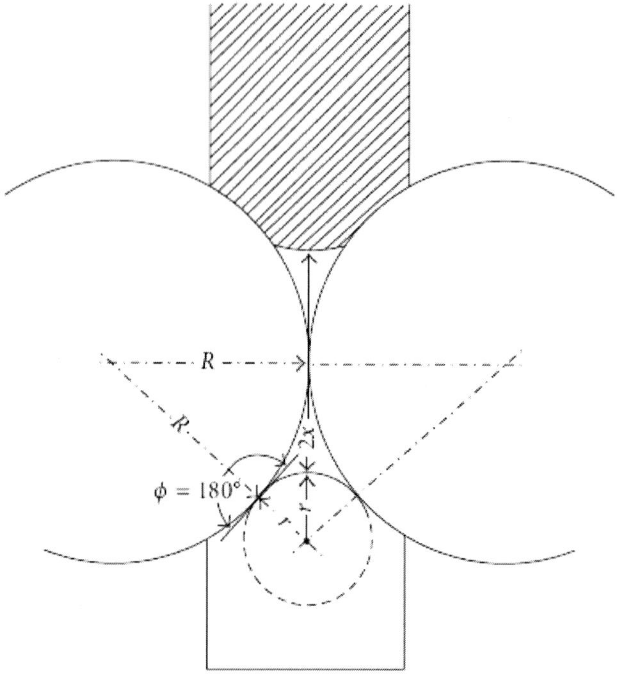


**Figure 4.** Measurement of mould filling ability.

$$R^2 + (r + x)^2 = (r + R)^2,$$

where $R$ = radius of the sand core, mm,

$R$ = 25 mm (radius of cylindrical core).

$r$ = radius of the meniscus ($2r = d$), mm,

$2x$ = distance between edges, mm,

$$\text{so } \frac{1}{d} = \frac{(R - x)}{x^2},$$

$$\frac{1}{d} = \text{mould filling ability}, \frac{1}{\text{mm}}, \tag{1}$$

Height gauge has been used to mark the height at every interval of 5 mm for each solidified casting. The distance between the fin edges, 2x, is measured with vernier micrometer for each 5 mm interval. The 2x values for 8 experiments are given at Table 4. The x values are substituted in (1) and the mould filling ability calculated for all the 8 experiments. The mould-filling ability values have been calculated for every 5 mm increment by using (1).

The mould filling ability values at various pressure heads for experimental run order 8 are given in Table 5. The mould filling ability for 8 experimental run orders has been given in Table 6.

## 4.1. Influence of Pressure Head on Mould Filling Ability of Aluminium Alloys

The mould filling ability is expressed as the reciprocal of the diameter of the meniscus of fin edge in the solidified test casting. It is expressed as a curve showing its variation with respect to pressure head (H) as measured on the test casting. So in present study the mould filling ability can be expressed as a curve showing its variation with pressure head. The mould filling ability values have been calculated for every 5 mm increment by using (1). The mould filling ability values at various pressure heads for experimental run orders 1 to 4 are calculated and plotted (Figure 5) showing that effect of pressure head on mould filling ability of LM25. This indicates that the mould filling ability increases with increase in pressure head.

The mould filling ability expressed as reciprocal of the diameter of the meniscus of fin edge in the solidified test casting. The mould filling ability is also plotted as a curve showing its variation with pressure head. Pressure head plays a very important role in the mould filling ability of metals. Experiments have been conducted to calculate mould filling ability of Aluminium alloys (LM6 and LM25). The mould filling ability increases with increase in pressure head. A higher-pressure head gives higher velocity of molten metal and better filling. Increase in pressure head leads to increase in the metallostatic force, allowing the liquid to enter the finer contours between the cylindrical cores thereby reducing the fin edge and increasing the mould filling ability values.

**Table 4.** The distance between the fin edges, 2 in mm.

| $H$, mm | Exp. run order 1 | Exp. run order 2 | Exp run order 3 | Exp run order 4 | Exp run order 5 | Exp run order 6 | Exp run order 7 | Exp run order 8 |
|---|---|---|---|---|---|---|---|---|
| 35 | 5.4 | | 7 | | 9.6 | 9.759 | 10.6 | 7.5 |
| 40 | 5.9 | 3.8 | 7.3 | 3.5 | 9.8 | 9.79 | 11 | 8 |
| 45 | 5.7 | 4 | 7.4 | 4.5 | 10 | 9.988 | 11.1 | 8.1 |
| 50 | 6 | 4.5 | 7.5 | 4.6 | 10.1 | 10.252 | 11.2 | 8.2 |
| 55 | 6 | 5.5 | 7.8 | 5 | 10.2 | 10.35 | 11.8 | 8.3 |
| 60 | 6.7 | 5.5 | 7.9 | 5.5 | 10.5 | 10.45 | 12 | 8.5 |
| 65 | 6.8 | 5.55 | 8 | 6 | 10.55 | 10.65 | 12.5 | 8.5 |
| 70 | 7.6 | 5.6 | 8.5 | 6.5 | 10.6 | 10.75 | 13 | 8.5 |
| 75 | 7.3 | 5.65 | 8.5 | 6.8 | 10.7 | 10.85 | 13.5 | 8.6 |
| 80 | 7.6 | 5.68 | 9 | 6.9 | 10.9 | 10.98 | 13.6 | 8.7 |
| 85 | 8.2 | 5.7 | 10 | 7 | 11 | 11.5 | 14 | 9 |
| 90 | 7.3 | 6 | 11 | 7.1 | | 11.55 | 14.1 | 10 |

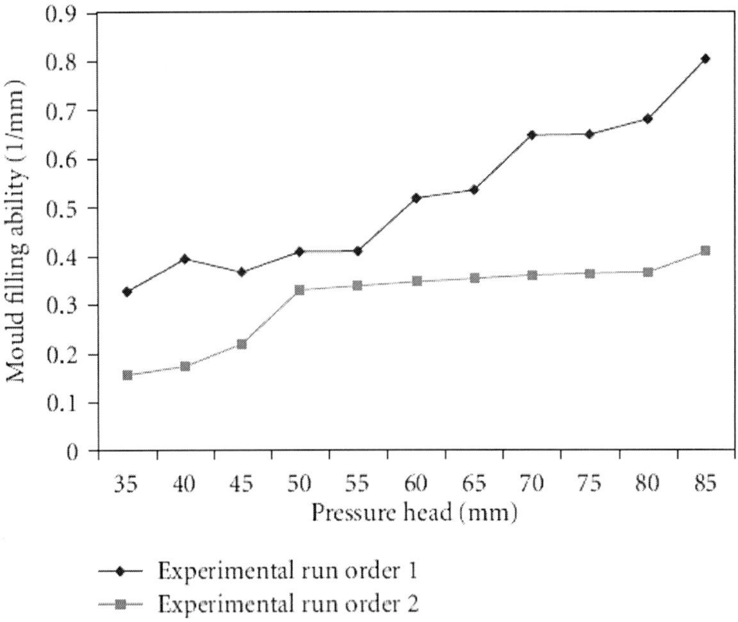


**Figure 5.** Influence of pressure head on mould filling ability.

If the moulds were at the same temperature as of the molten metal during pouring, the mould filling ability follows the equation given below:

$$\frac{1}{d} = \frac{h.\rho.g}{2\sigma \cdot \cos\varphi},$$

$\frac{1}{d}$ = mould filling ability, 1/mm,

$h$ = pressure head, mm,

$\rho$ = density, gm/cc,

$g$ = acceleration due to gravity, m/sec$^2$,

$\sigma$ = surface tension, dynes/cm,

$\varphi$ = angle of contact. (2)

**Table 5**. Mould filling ability calculations.

| S. no. | $H$, mm | $2x$, mm | $1/d$ mm$^{-1}$, $[(R-x)/x^2]$ |
|---|---|---|---|
| 1 | 0 | 5 | 0.278 |
| 2 | 5 | 5.5 | 0.34 |
| 3 | 10 | 6 | 0.408 |
| 4 | 15 | 6.2 | 0.438 |
| 5 | 20 | 6.5 | 0.485 |
| 6 | 25 | 6.8 | 0.536 |
| 7 | 30 | 7 | 0.569 |
| 8 | 35 | 7.5 | 0.662 |
| 9 | 40 | 8 | 0.762 |
| 10 | 45 | 8.1 | 0.783 |
| 11 | 50 | 8.2 | 0.806 |
| 12 | 55 | 8.3 | 0.826 |
| 13 | 60 | 8.5 | 0.871 |
| 14 | 65 | 8.5 | 0.871 |
| 15 | 70 | 8.5 | 0.871 |
| 16 | 75 | 8.6 | 0.892 |
| 17 | 80 | 8.7 | 0.9165 |
| 18 | 85 | 9 | 0.99 |
| 19 | 90 | 10 | 1.25 |

The factors which influence the mould filling ability in liquid region are surface tension and contact angle. The factors which influence the mould filling ability in solidifying region are cooling conditions. In the present studies the experiments have been conducted in solidifying region. So the mould filling ability depends on pressure head while the surface tension and density are neglected and an angle of contact of 180° is assumed. Higher pressure heads lead to higher mould filling ability because the metal has to be in contact with larger area in order to penetrate into cavity which promotes faster solidification.

**Table 6.** The mould-filling ability for 8 experimental run orders.

| *H*, mm | Exp. run order 1 | Exp. run order 2 | Exp. run order 3 | Exp. run order 4 | Exp. run order 5 | Exp. run order 6 | Exp. run order 7 | Exp. run order 8 |
|---|---|---|---|---|---|---|---|---|
| 35 | 0.326906 | 0.14784 | 0.569767 | 0.124 | 1.14 | 1.1833 | 1.425888 | 0.6617 |
| 40 | 0.394671 | 0.156277 | 0.624005 | 0.131 | 1.19 | 1.191794 | 1.551282 | 0.7619 |
| 45 | 0.366704 | 0.17391 | 0.642723 | 0.22 | 0.125125 | 1.2466 | 1.583676 | 0.7829 |
| 50 | 0.409091 | 0.22 | 0.661765 | 0.233 | 1.2783 | 1.322 | 1.616495 | 0.8043 |
| 55 | 0.41 | 0.33 | 0.720853 | 0.2777 | 1.307 | 1.35 | 1.822513 | 0.826 |
| 60 | 0.51836 | 0.338 | 0.741211 | 0.33988 | 1.39577 | 1.38 | 1.894737 | 0.86 |
| 65 | 0.535185 | 0.3464 | 0.761905 | 0.409 | 1.41 | 1.44 | 2.083333 | 0.87 |
| 70 | 0.648113 | 0.3531 | 0.870482 | 0.48565 | 1.42588 | 1.47 | 2.283784 | 0.88 |
| 75 | 0.65 | 0.3598 | 0.881815 | 0.535 | 1.4566 | 1.503 | 2.496575 | 0.89 |
| 80 | 0.681132 | 0.3639 | 0.987805 | 0.55 | 1.5193 | 1.544 | 2.540659 | 0.9136 |
| 85 | 0.8043 | 0.3666 | 1.25 | 0.5697 | 1.5512 | 1.7175 | 2.722222 | 0.98 |
| 90 | 0.826 | 0.4091 | 1.551282 | 0.587529 | 1.583676 | 1.735 | 2.768942 | 1.25 |

## 4.2. Influence of Silicon Content on Mould Filling Ability

The mould filling ability increases with increase in silicon content in the alloy. LM6 is the eutectic alloy and it contains more silicon content of 12 weight percentage. Higher the silicon content higher will be the fluidity. Silicon additions improve casting characteristics by improving fluidity, feeding, and hot tear resistance.

This has been shown in Figure 6. LM6 having better mould filling ability than LM25 alloy attributes mainly to the higher silicon content in LM6 alloy. The mould filling ability values corresponding to Figure 6 have been calculated using (1). The mould filling ability values at various pressure heads for LM6 and LM25 are plotted.

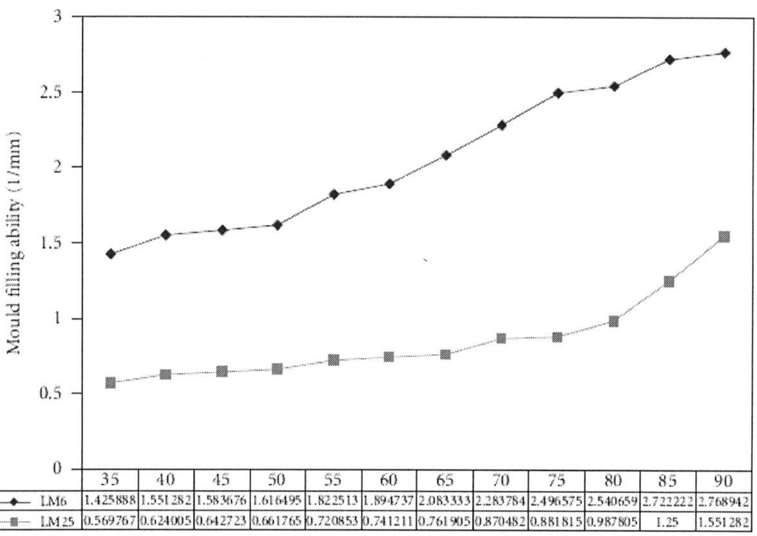


| | 35 | 40 | 45 | 50 | 55 | 60 | 65 | 70 | 75 | 80 | 85 | 90 |
|---|---|---|---|---|---|---|---|---|---|---|---|---|
| LM6 | 1.425888 | 1.551282 | 1.583676 | 1.616495 | 1.822513 | 1.894737 | 2.083333 | 2.283784 | 2.496575 | 2.540659 | 2.722222 | 2.768942 |
| LM25 | 0.569767 | 0.624005 | 0.642723 | 0.661765 | 0.720853 | 0.741211 | 0.761905 | 0.870482 | 0.881815 | 0.987805 | 1.25 | 1.551282 |

Pressure head (mm)

**Figure 6.** Influence of silicon on mould filling ability of LM6 and LM25.

## 4.3. Influence of Pouring Temperature on Mould Filling Ability

Pouring temperature affects the mould filling ability. Mould filling ability value increases with increase in pouring temperature. From the present experiments, it has been shown in Figure 7. Increase in pouring temperature increases the heat content of the alloy, resulting in alloy being liquid for longer duration. This promotes the metallostatic pressure head to act on the liquid metal for a longer period of time, thus forcing the metal to easily enter the fine contours.

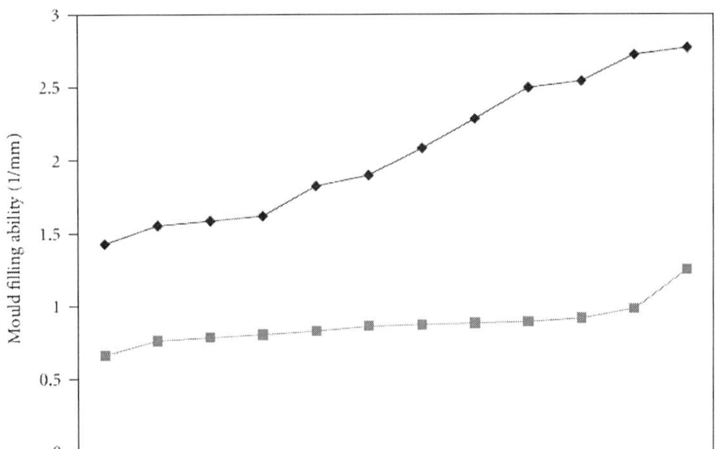


| | 35 | 40 | 45 | 50 | 55 | 60 | 65 | 70 | 75 | 80 | 85 | 90 |
|---|---|---|---|---|---|---|---|---|---|---|---|---|
| LM6 *T* + 20 | 1.425888 | 1.551282 | 1.583676 | 1.616495 | 1.822513 | 1.894737 | 2.083333 | 2.283784 | 2.496575 | 2.540659 | 2.722222 | 2.768942 |
| LM6 *T* | 0.6617 | 0.7619 | 0.7829 | 0.8043 | 0.826 | 0.86 | 0.87 | 0.88 | 0.89 | 0.9136 | 0.98 | 1.25 |

Pressure head (mm)

**Figure 7.** Influence of pouring temperature on mould filling ability of LM6.

## 4.4. Influence of Sand Fineness on Mould Filling Ability

Sand fineness plays a very important role in the moulding process. In the following experiments, the fineness of sand being used for cylindrical cores varies from fine sand and to coarse sand. Experimental results show that the mould filling ability increases with increase in sand fineness from Figure 8. The coarse sand is having more chilling power and solidifies the metal faster. This restricts the filling up of the cavity contours between the cylindrical sand cores and reduces the mould filling ability.

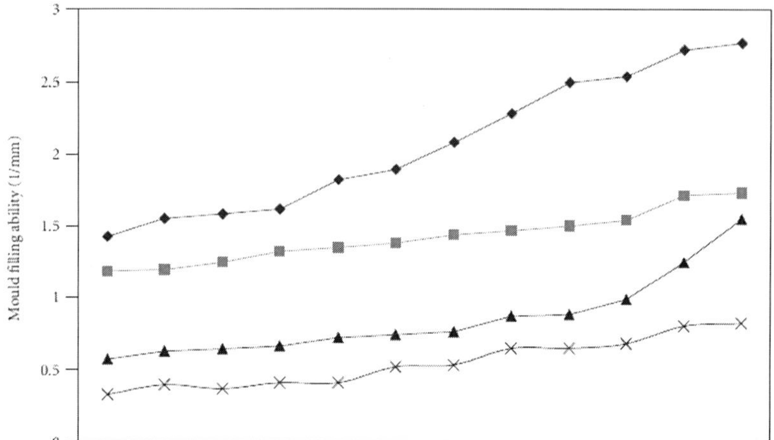


| | 35 | 40 | 45 | 50 | 55 | 60 | 65 | 70 | 75 | 80 | 85 | 90 |
|---|---|---|---|---|---|---|---|---|---|---|---|---|
| LM6AFS40 | 1.425888 | 1.551282 | 1.583676 | 1.616495 | 1.822513 | 1.894737 | 2.083333 | 2.283784 | 2.496575 | 2.540659 | 2.722222 | 2.768942 |
| LM6AFS25 | 1.1833 | 1.191794 | 1.2466 | 1.322 | 1.35 | 1.38 | 1.44 | 1.47 | 1.503 | 1.544 | 1.7175 | 1.735 |
| LM25AFS40 | 0.569767 | 0.624005 | 0.642723 | 0.661765 | 0.720853 | 0.741211 | 0.761905 | 0.870482 | 0.881815 | 0.987805 | 1.25 | 1.551282 |
| LM25AFS25 | 0.326906 | 0.394671 | 0.366704 | 0.409091 | 0.41 | 0.51836 | 0.535185 | 0.648113 | 0.65 | 0.681132 | 0.8043 | 0.826 |

Pressure head (mm)

**Figure 8.** Influence of fineness of sand on mould filling ability of LM6 and LM25.

## 4.5. Influence of Mould Coating on Mould Filling Ability

Mould coating plays a significant role in the moulding process. Experiments have been conducted with graphite coating on the mould. Mould coating improves the mould filling ability. This is by creating a smooth mould metal interface which provides less resistance to metal filling the contours. Results are shown in Figure 9.

## 4.6. Analysis of Variance (ANOVA)

To study the characteristics of mould filling ability, it is necessary to study, all the process parameters together. In the present study Taguchi orthogonal array L8 (2**4) is used with four factors and 2 levels (for 8 runs) as indicated in Table 2 and response values are given in Table 7. The interactions selected for this study are(i)alloy and sand fineness (AFS no.),(ii)alloy and mould coat,(iii)alloy and pouring temperature.

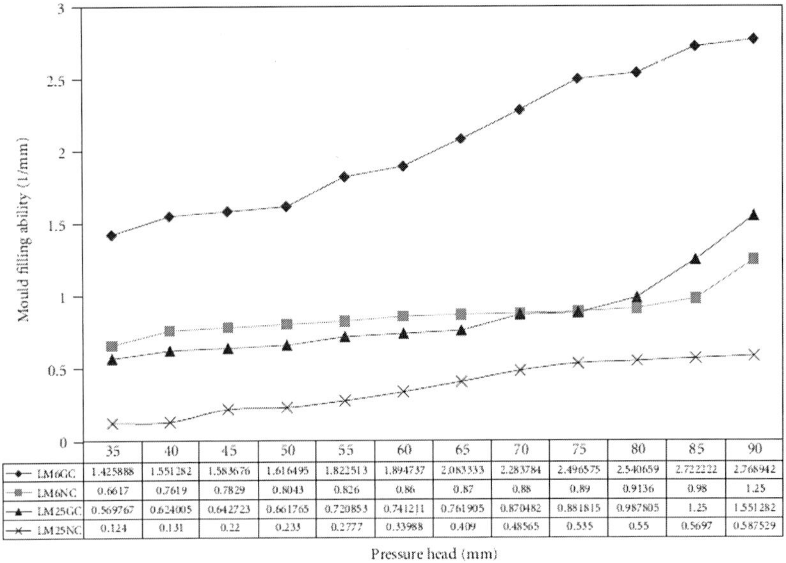


| | 35 | 40 | 45 | 50 | 55 | 60 | 65 | 70 | 75 | 80 | 85 | 90 |
|---|---|---|---|---|---|---|---|---|---|---|---|---|
| LM6GC | 1.425888 | 1.551282 | 1.583676 | 1.616495 | 1.822513 | 1.894737 | 2.083333 | 2.283784 | 2.496575 | 2.540659 | 2.722222 | 2.768942 |
| LM6NC | 0.6617 | 0.7619 | 0.7829 | 0.8043 | 0.826 | 0.86 | 0.87 | 0.88 | 0.89 | 0.9136 | 0.98 | 1.25 |
| LM25GC | 0.569767 | 0.624005 | 0.642723 | 0.661765 | 0.720853 | 0.741211 | 0.761905 | 0.870482 | 0.881815 | 0.987805 | 1.25 | 1.551282 |
| LM25NC | 0.124 | 0.131 | 0.22 | 0.233 | 0.2777 | 0.33988 | 0.409 | 0.48565 | 0.535 | 0.55 | 0.5697 | 0.587529 |

**Figure 9.** Influence of graphite coating on mould filling ability of LM6 and LM25.

ANOVA values for mould filling ability are calculated using MINITAB and it is given at Table 8.

P is used to determine whether a factor is significant, typically to compare against an alpha () value of 0.1. Alpha is frequently referred to as the level of significance and is expressed as a probability ranging between 0 and 1. If the P value is lower than 0.1, then the factor is significant. Choosing a larger value for  is better for aerospace applications. The analysis of variance indicates that the alloy factor is considered more significant in influencing mould filling ability. It is seen that the alloy content and pouring temperature interaction contribute more to mould filling ability value.

Use of main effects' plot in conjunction with an analysis of variance (ANOVA) is useful when there have been several factors. The effect is the mean for that level minus the overall mean for the factor. The effects are summarized at Table 9. The overall mean is 1.309 for the present study from the ANVOA calculations.

**Table 7.** Response table.

| Exp. run order | Alloy | Sand fineness (AFS no.) | Mould coat | Pouring temperature (°C) | $1/d$ mm$^{-1}$ |
|---|---|---|---|---|---|
| 1 | LM25 | 25 | Graphite (GC) | $T$ | 0.624005 |
| 2 | LM25 | 25 | No coating (NC) | $T + 20$ | 0.40909 |
| 3 | LM25 | 40 | Graphite (GC) | $T$ | 1.55 |
| 4 | LM25 | 40 | No coating (NC) | $T + 20$ | 0.58753 |
| 5 | LM6 | 25 | No coating (NC) | $T$ | 1.55 |
| 6 | LM6 | 25 | Graphite (GC) | $T + 20$ | 1.73453 |
| 7 | LM6 | 40 | Graphite (GC) | $T + 20$ | 2.7689 |
| 8 | LM6 | 40 | No coating (NC) | $T$ | 1.25 |

**Table 8.** ANOVA.

| Source | Degrees of freedom | Sum of squares | Mean squares | Factor | *P* |
|---|---|---|---|---|---|
| Alloy | 1 | 2.13501 | 2.13501 | 8.33 | 0.045 |
| AFS no. | 1 | 0.42265 | 0.42265 | 1.02 | 0.370 |
| Pouring temperature | 1 | 0.03459 | 0.03459 | 0.14 | 0.732 |
| Mould coat | 1 | 1.03739 | 1.03739 | 4.05 | 0.114 |
| Alloy * sand fineness | 1 | 0.01712 | 0.01712 | 0.04 | 0.849 |
| Alloy * mould coat | 1 | 0.03459 | 0.03459 | 0.14 | 0.732 |
| Alloy * pouring temperature | 1 | 1.03739 | 1.03739 | 4.05 | 0.114 |
| Error | 4 | 1.02461 | 0.25615 | | |
| Total | 11 | 5.74335 | | | |

**Table 9.** The effects are summarized.

| Factor | Mean | Effect (mean – overall mean) | |
|---|---|---|---|
| LM6 | 1.82586 | 1.82586 – 1.309 | 0.51686 |
| LM25 | 0.79266 | 0.79266 – 1.309 | −0.51634 |
| Pouring temp *T* | 1.24350 | 1.24350 – 1.309 | −0.0655 |
| Pouring temp *T* + 20 | 1.37501 | 1.37501 – 1.309 | 0.06601 |
| GC | 1.66936 | 1.66936 – 1.309 | 0.36036 |
| NC | 0.94915 | 0.94915 – 1.309 | −0.35985 |
| 25 | 1.07941 | 1.07941 – 1.309 | −0.22959 |
| 40 | 1.53911 | 1.53911 – 1.309 | 0.23011 |

Main effects' plot for mould filling ability values isshown in Figure 10. The magnitude of an effect is related to its distance from the mean. A reference line is drawn for the representing the overall mean. Lines are not horizontal (parallel to the x-axis), for all the four factors of the present study, so there is a main effect present. The main effects plot indicates that the alloy, mould coat, sand fineness, and pouring temperature are influencing the mould filling ability. In case of alloy factor, the difference in the vertical position of the plotted points is more, hence magnitude of the main effect is greater.

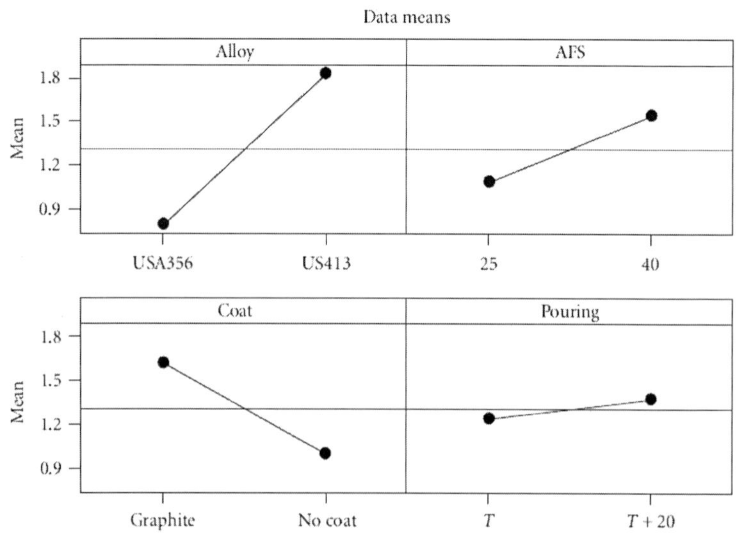


**Figure 10.** Main effects' plot for mould filling ability.

The main effects' plot indicates an optimum factor combination of LM6, graphite coat, sand fineness no. 40, and pouring temperature for getting better mould filling ability.

Correlation coefficient quantifies the strength of the linear association between two variables.

To ascertain the strength of association, correlation co-efficient between the two alloys is calculated using the formula

$$\text{Correlation}(r) = \left[\frac{(N\Sigma XY - (\Sigma X)(\Sigma Y))}{\text{Sqrt}\left(\left[N\Sigma X^2 - (\Sigma X)^2\right]\left[N\Sigma Y^2 - (\Sigma Y)^2\right]\right)}\right], \quad (3)$$

Where N = number of values, X = mould filling ability values for LM6, and Y = mould filling ability values for LM 25.

The results of mould filling ability values used to find the correlation coefficient value are given in Table 10.

**Table 10.** Correlation coefficient for mould filling ability values.

| $N$ | $X$ (LM6) | $Y$ (LM 25) | $X \times Y$ | $X^2$ | $Y^2$ |
|---|---|---|---|---|---|
| 1 | 1.55 | 0.624 | 0.967208 | 2.4025 | 0.389382 |
| 2 | 1.7345 | 0.4091 | 0.709579 | 3.008594 | 0.167355 |
| 3 | 2.7689 | 1.55 | 4.291795 | 7.666807 | 2.4025 |
| 4 | 1.25 | 0.5875 | 0.734413 | 1.5625 | 0.345192 |

The correlation coefficient always takes a value between −1 and 1, with +1 indicating perfect correlation.

The correlation coefficient value obtained is 0.899242 indicating a strong, positive relationship.

## 5. CONCLUSIONS

The mould filling ability characteristics of LM6 and LM25 alloys have been studied using pin test piece with cylindrical cores. Mould filling ability increases with increase in metallostatic/pressure head. Silicon content in the alloys improves the mould filling ability. LM6 is having better mould filling ability than LM25. The mould filling ability increases with increase in fineness of sand because coarse sand is having more chilling power and solidifies the metal faster. Mould coat improves mould filling ability by providing less resistance to liquid metal flow. Alloy factor has been considered more significant in influencing mould filling ability. The main effects' plot has been indicating an optimum factor combination of LM6, graphite coat, sand fineness no. 40, and pouring temperature  for getting better mould filling ability.

## ACKNOWLEDGMENTS

The authors extend their thanks to the Director, the Directorate of Engineering, and the Director DRDL for providing support and permission for carrying out this R&D work.

## REFERENCES

1. ASM Metals Handbook, Volume 15, Casting, ASM INTERNATIONAL, The Materials Information Company.

2. S. V. Shepel and S. Paolucci, "Numerical simulation of filling and solidification of permanent mold castings," Applied Thermal Engineering, vol. 22, no. 2, pp. 229–248, 2002.

3. Prepared by Energetics, Incorporated: Aluminum industry roadmap for the automotive market: Enabling Technologies and Challenges for Body Structures and Closures, The Aluminum Association, Inc., with support from the U.S. Department of Energy, 1999.

4. Y. Su, T. Zhang, J. Guo et al., "Physical simulation of mold-filling processing of thin-walled castings under traveling magnetic field,"

Journal of Materials Science and Technology, vol. 20, no. 1, pp. 27–30, 2004.

5. S. Sundarrajan and H. M. Roshan, "Studies on mould filling ability characteristics of Mg-Al Alloys,"Transactions of the American Foundrymen's Society, pp. 607–616, 1989.

6. R. W. Lewis, A. S. Usmani, and J. T. Cross, "Efficient mould filling simulation in castings by an explicit finite element method," International Journal for Numerical Methods in Fluids, vol. 20, no. 6, pp. 493–506, 1995.

7. M. C. Flemings, Solidification Processing, Mcgraw-Hill, London, UK, 1974.

8. N. R. Green and J. Campbell, "Influence in oxide film filling defects on the strength of Al-7si-Mg Alloy Castings," Transactions of the American Foundry Society, vol. 114, pp. 341–347, 1994.

9. C. Reilly, N. R. Green, M. R. Jolly, and J. C. Gebelin, "Using the calculated froude number for quality assessment of casting filling methods," in Proceedings of the 12th International Conference on Modeling of Casting, Welding, and Advanced Solidification Processes, pp. 419–426, June 2009.

10. M. Jolly, "Prof. John Campbell's ten rules for making reliable castings," JOM, vol. 57, no. 5, pp. 19–28, 2005.

11. C. J. Cooksey, V. Kondic, and J. Wilcock, "The casting fluidity of some foundry alloys," The British Foundryman, vol. 52, p. 381, 1959.

12. B. Sirrell and J. Campbell, "Mechanism of filtration in reduction of casting defects due to surface turbulence during mould filling," Transactions of the American Foundrymen's Society, vol. 105, pp. 645–654, 1998.

13. User Manual of MINITAB.

14. D. Hanumantha Rao, G. R. N. Tagore, and G. Rangajanardhan, "Process simulation of aluminium alloy casting solidification for controlled porosity and cooling rate," Indian Foundry Journal, vol. 52, no. 5, pp. 26–34, 2006.

12

# PREPARATION OF METALLIC IRON POWDER FROM PYRITE CINDER BY CARBOTHERMIC REDUCTION AND MAGNETIC SEPARATION

**Hongming Long, Tiejun Chun *, Zhanxia Di, Ping Wang, Qingmin Meng and Jiaxin Li**

School of Metallurgical Engineering, Anhui University of Technology, Ma'anshan 243002, China

## ABSTRACT

The reduction and magnetic separation procedure of pyrite cinder in the presence of a borax additive was performed for the preparation of reduced powder. The effects of borax dosage, reduction temperature, reduction time and grinding fineness were investigated. The results show that when pyrite cinder briquettes with 5% borax were pre-oxidized at 1050 °C for 10 min, and reduced at 1050 °C for 80 min, with the grinding fineness (<0.44 mm) passing 81%, the iron recovery was 91.71% and the iron grade of the magnetic concentrate was 92.98%. In addition, the microstructures of the products were analyzed by optical microscope, scanning electron microscope (SEM), and mineralography, and the products were also studied by the X-ray powder diffraction technique (XRD) to investigate the mechanism; the results show that the borax additive was approved as a good additive to improve the separation of iron and gangue.

## KEYWORDS

pyrite cinder; carbothermic reduction; magnetic separation; metallic iron powder

## 1. INTRODUCTION

Pyrite cinders are produced as a by-product of the sulfuric acid industry, containing $Fe_2O_3$, $Fe_3O_4$, metallic silicate, hazardous heavy metals, *etc.* In China, more than 30% of sulfuric acid is produced from pyrites ores, and more than 10 million tons of pyrite cinders are produced annually, with the total generation of pyrite cinders at over 10 million tons per [1,2]. The storage of pyrite cinder consumes a great deal of land and creates a serious environmental problem [3,4]. Nevertheless, the utilization of pyrite cinders was greatly limited due to its low iron grade and high sulphur content [5].

In the previous research, physical separation processes have been used to separate gangue and iron, including gravity separation, magnetic separation, and floatation [6,7,8]. However, as the high-temperature procedure alters the physical and chemical properties of pyrite cinder, it was hard to separate the iron and gangue of pyrite cinder by transitional physical methods, and the iron grade of the iron concentrate obtained was also low. In addition, chemical methods [9,10] were also used to treat the pyrite cinder. Nevertheless, the chemical methods have a high cost of reducing reagents which could lead to secondary pollution. Based on the aforementioned introduction, it can be found that a capable method to separate the iron and gangue of the pyrite cinder efficiently is significant for realizing the utilization and preparing the iron concentrate from pyrite cinder.

Therefore, this study will focus on the recovery of the iron and the preparation of reduced powder obtained from pyrite cinder by adopting the reduction-magnetic separation method which has been proved to be an effective way to recover iron from solid wastes [11,12]. The effects of additive dosage, reduction temperature, reduction time and grinding fineness on the iron recovery are investigated.

# 2. MATERIALS AND METHODS

## 2.1. Raw Material

Table 1 shows the chemical composition of pyrite cinder. The contents of Fe and FeO are 60.95% and 12.61%. The contents of $SiO_2$, $Al_2O_3$, CaO and MgO are 5.92%, 1.98%, 0.47% and 0.36%, respectively. The particle size of pyrite cinder can be found in Table 2. The analytical reagent of borax ($Na_2B_4O_7 \cdot 10H_2O$) was used during the experiment.

**Table 1.** Chemical composition of pyrite cinder (mass/%). TFe: Total Fe; LOI: Loss of ignition.

| Element | TFe | FeO | $SiO_2$ | $Al_2O_3$ | CaO | MgO | $K_2O$ | $Na_2O$ | P | S | LOI |
|---|---|---|---|---|---|---|---|---|---|---|---|
| Content | 60.95 | 12.61 | 5.92 | 1.98 | 0.47 | 0.36 | 0.21 | 0.05 | 0.04 | 0.39 | 1.12 |

**Table 2.** Particle size of pyrite cinder (mass/%).

| Size | 0.15–0.106 mm | 0.106–0.074 mm | 0.074–0.044 mm | 0.044 mm |
|---|---|---|---|---|
| Content | 6.47 | 14.97 | 19.85 | 58.71 |

Bituminous coal was used as a reductant. The industrial analysis results and particle size are given in Table 3 and Table 4. The fixed carbon is 50.04%, volatile is 30.15% and ash content is 12.04%, which is suitable for the reduction [13]. The particle size of bituminous coal is 100% passing through 5 mm and 75.52% passing through 0.5 mm.

**Table 3.** Industrial analysis results of bituminous coal (mass/%).

| Item | Fixed Carbon | Moisture | Ash | Volatiles | Sulfur |
|---|---|---|---|---|---|
| Content | 50.04 | 7.97 | 12.04 | 30.15 | 0.37 |

**Table 4.** Particle size of bituminous coal (mass/%).

| Size | 5–1 mm | 1–0.5 mm | 0.5–0 mm |
|---|---|---|---|
| Content | 58.29 | 17.23 | 24.48 |

## 2.2. Methods

### 2.2.1. Reduction Roasting and Magnetic Separation

The main process of the experiment contains five steps, which are briquetting, pre-oxidizing, reduction, ball-milling and magnetic separation. Firstly, the pyrite cinder and borax additives were mixed to prepare the briquettes. The briquettes were made by pressing the material mixture in a cylindrical mould for 1 min under the pressure of 8 MPa. The briquettes made are 10 mm in diameter and 15 mm in height. The drying of the briquettes was carried out in the oven at 105 °C for 4 h.

The pre-oxidization of pyrite cinder briquettes was carried out in a tube furnace (Φ 50 × 600 mm, Brother company, Zhengzhou, China). The briquettes were loaded in a porcelain boat crucible of 60 mm in length and 30 mm in width. Then the porcelain boat was put into the central area of the tube furnace and oxidized at 1050 °C for 10 min.

After the pre-oxidization, the pre-oxidized briquettes were reduced in a shaft furnace (Φ 80 × 800 mm, Brother company, Zhengzhou, China). The graphite crucible of 60 mm in diameter and 150 mm in height, containing 30 g pre-oxidized briquettes and 60 g reductant, was put into the shaft furnace after the target temperature was reached. The upper end of corundum tube was covered with a refractory block during reduction in order to maintain enough CO concentration in the shaft furnace. After the reduction, nitrogen with a flow rate of 1 L/min was introduced to the crucible to cool down to prevent deoxidizing. Through that, the reduced pyrite cinder briquettes were obtained.

The magnetic separation was performed in the magnetic tube after the reduced briquettes were ground in the cone ball mill. The final metallic iron powder was obtained after magnetic separation at the magnetic intensity of 0.08 T. The model of the cone ball mill is XMQ240 × 90 (Luoke company, Wuhan, China) and the laboratory magnetic equipment is a Davis tube of XCGS-73 (Shunze company, Changsha, China).

### 2.2.2. Methods of Analysis

The chemical compositions of samples and the final product were obtained from ICP-MS (inductively coupled plasma mass spectrometry, Thermo Electron corporation, Waltham, MA, America). The crystalline phase of the samples was investigated by the XRD technique, and the structures and element distribution of the reduced briquettes were analyzed by scanning electron microscope (SEM) (Oxford, UK) with energy dispersion system (EDS) (Oxford, UK).

The total iron (TFe, wt. %) content and metallic iron (MFe, wt. %) content in the iron concentrate (metallic iron powder) were obtained by chemical methods, and the metallization degree, η, was calculated by

$$\eta = \frac{MFe}{TFe} \times 100\% \qquad (1)$$

The iron recovery, ε, was calculated by the following formula:

$$\varepsilon = \gamma \times \frac{\beta}{\alpha} \qquad (2)$$

where $\gamma$ is the yield of iron concentrate during the magnetic separation, wt. %; $\beta$ is the Fe content of the iron concentrate, wt. %; $\alpha$ is the Fe content of the reduced briquettes, wt. %.

## 3. RESULTS AND DISCUSSION

### 3.1. Effect of Borax Additive Dosage

The effects of the dosage of borax on the iron grade of the magnetic concentrate, iron recovery and metallization degree of the reduced briquettes were investigated, which was shown in Figure 1. With the increase of the borax dosage, the iron grade of the magnetic concentrate increases, but the iron recovery decreases (Figure 1a). It is mainly because the borax can enhance the lattice defect of the iron ore [14]. During the pre-oxidization, the borax can improve the re-crystallization of the iron oxide crystal grains and therefore enlarge the iron oxide particles [15]. On the other hand, borax also creates more lattice defects of metallic iron to improve the re-crystallization of the metallic iron crystal grains, which leads to the growth and interconnection of iron particles (Figure 2 and Figure 3). The remarkable interconnection and growth of iron particles promote the dissociation of iron and gangue, so the iron grade goes up. As the interconnection and growth of iron particles cause weak diffusion of CO in the briquettes, it imposes a negative influence on the metallization degree (Figure 1b). The low metallization degree results in the decrease in the iron recovery rate (Figure 1b). When the dosage of the borax exceeds 5%, the iron grade of the magnetic concentrate declines. It can be explained by the fact that the borax is liquid during the test temperature range, and the amount of liquid borax limits the interconnection and growth of iron particles (Figure 2), which leads to the decrease of the iron grade of the magnetic concentrate.

## 3.2. Effect of Reduction Temperature

The effect of the reduction temperature on the metallization degree of the reduced briquettes, iron grade and iron recovery of the magnetic concentrate are shown in Figure 4. It is clear that the iron grade of the magnetic concentrate and the recovery rate of iron both increase with the increase in temperature, especially the iron recovery (Figure 4a). The reason is that the Boudouard reaction is very active at high temperatures, which improves the metallization degree of the reduced briquettes (Figure 4b). Furthermore, a high reduction temperature accelerates the re-crystallization of the crystal grains of metallic iron. The particle size of metallic iron is larger at higher temperatures (Figure 5), so the dissociation of iron and gangue becomes easier.

## 3.3. Effect of Reduction Time

Figure 6 displays the influence of reduction time on the metallization degree of the reduced briquettes, iron grade and iron recovery of the magnetic concentrate. The metallization degree of the reduced briquettes significantly rises with the prolonged reduction time (Figure 6a), so both the iron grade and iron recovery of the magnetic concentrate increase (Figure 6b). Also, the particle size of metallic iron has a clear uptrend with a long reduction time [16].

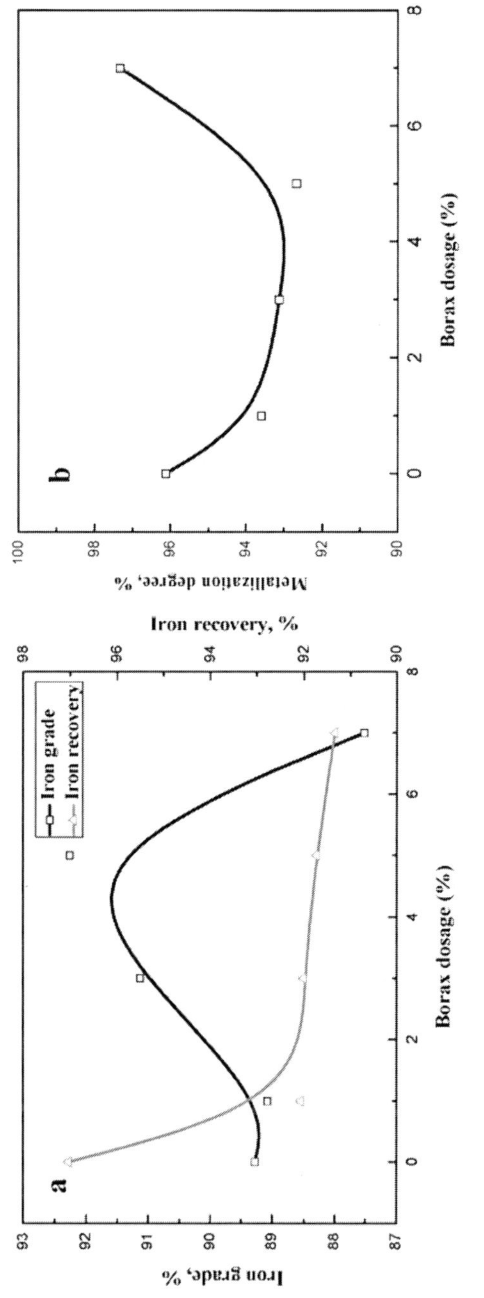


**Figure 1.** Effect of borax dosage on iron grade and iron recovery of magnetic concentrate (**a**); and effect of borax dosage on metallization degree of reduced briquettes (**b**) (reducing at 1050 °C for 80 min, and grinding fineness (<0.044 mm) exceeding 68%).

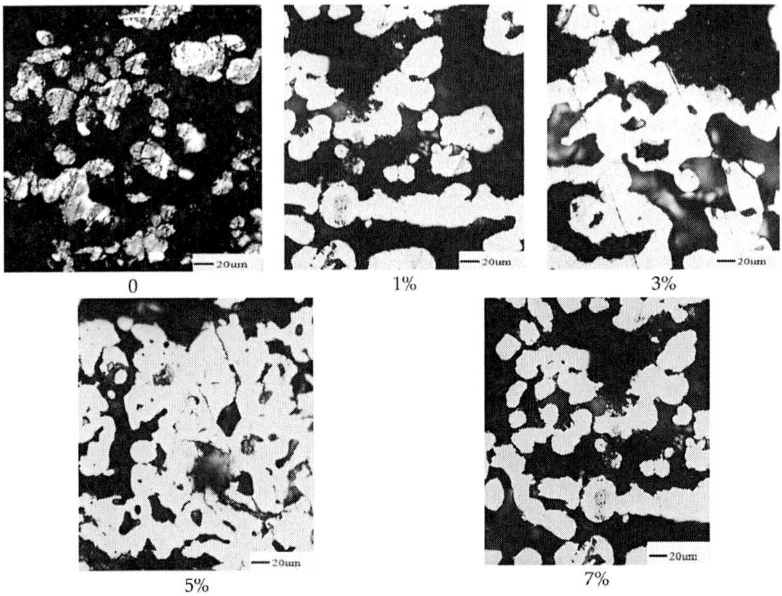


**Figure 2.** Microstructure of reduced briquettes with different dosage of borax (bright white, metallic iron).

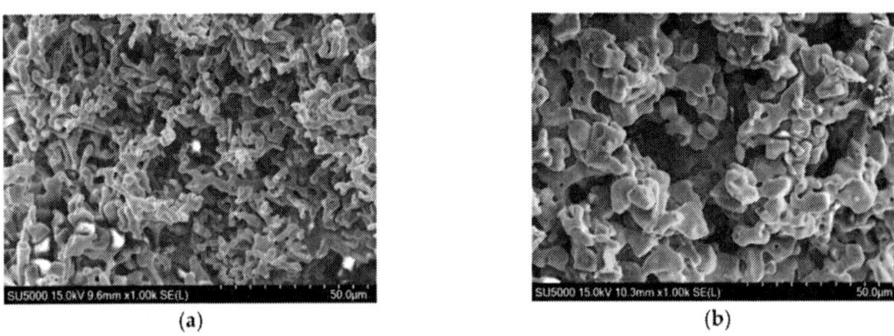


**Figure 3.** Microstructure of reduced briquettes with different dosage of borax by scanning electron microscope (SEM) ((**a**): 0% borax; (**b**): 5% borax).

## 3.4. Effect of Grinding Fineness

Figure 7 shows the effect of grinding fineness on the iron grade and iron recovery of the magnetic concentrate. It illustrates that the recovery rate of the magnetic concentrate drops with the increase of grinding fineness. The iron grade of the magnetic concentrate was promoted. All these can be explained by the fact that when the grinding particle is fine, the dissociation degree of the iron particles of metallic iron and the gangue particles is enhanced. When the grinding fineness (<44 μm) exceeds 81%, there is a drop in the iron grade of the magnetic concentrate. This is because the iron ore and gangue irons are completely mixed when the grinding particle is too fine, which weakens the separation efficiency of the iron and gangue.

## 3.5. Analysis of the Final Product

The magnetic concentrates achieved were analyzed by XRD and SEM-EDS. XRD patterns of the magnetic concentrate and pyrite cinder are displayed in Figure 8. The iron of the pyrite cinder before pre-oxidization is mainly in the form of magnetite, hematite and maghemite, and the gangue is in the form of quartzes. After the reduction, ball milling and magnetic separation, the magnetic concentration prepared mostly presents in metallic phase, and no gangue phase exists. Thus, the dissociation of iron and gangue executes well.

SEM-EDS of the magnetic concentrate under the optimum condition is shown in Figure 9. The particles of the magnetic concentration are mainly in the shape of flakes due to good ductility of the metallic iron. As shown in Figure 9, the particles such as point A are mainly in metallic iron phases. The particles such as point B are the gangue and some iron minerals. The gangue particle in the magnetic concentrate is embedded in the particle of the metallic iron, which indicates it is very difficult to remove the residual gangue in magnetic concentrate.

Compared with the traditional recovery of iron from iron ores, the operation cost of adding a borax reduction roasting to treat the pyrite cinder is high, but the technology of adding borax roasting and magnetic separation opens the way to recover iron from the pyrite cinder [17,18,19].

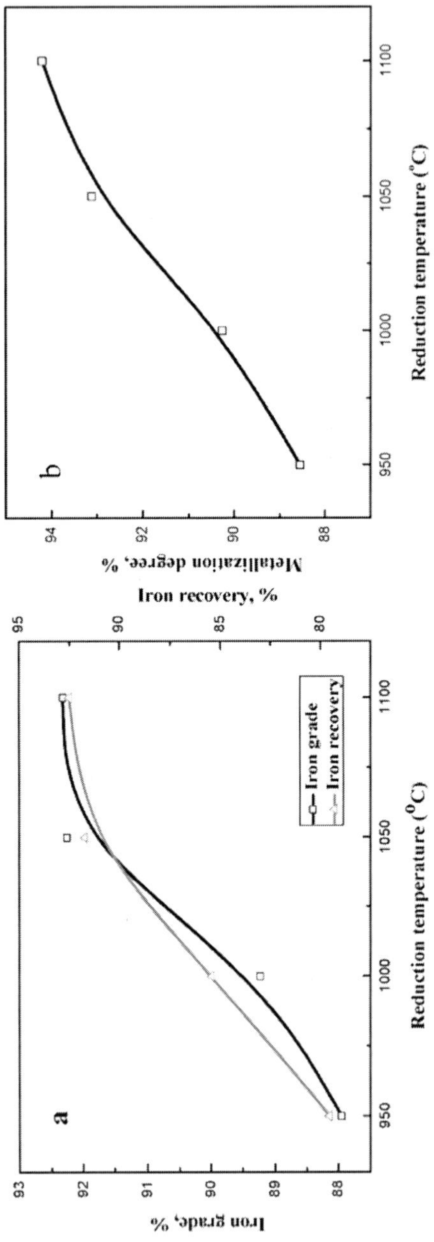


**Figure 4.** Effect of reduction temperature on iron grade and iron recovery of magnetic concentrate (**a**); and effect of reduction temperature on metallization degree of reduced briquettes (**b**) (borax dosage of 5%, reducing for 80 min, and grinding fineness (<0.044 mm) exceeding 68%).

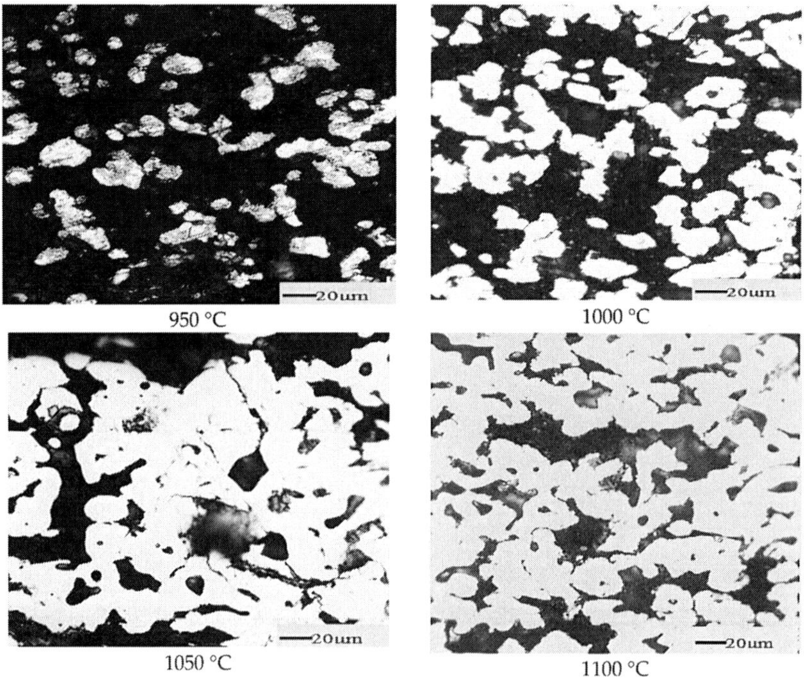


**Figure 5.** Microstructure of reduced briquettes at different reduction temperature (bright white, metallic iron).

# 4. CONCLUSIONS

Reduction roasting and magnetic separation is one of the efficient ways to retrieve the metallic iron powder from pyrite cinder, and borax is an effective additive to improve the separation efficiency of the iron and gangue of the pyrite cinder. During the reduction process, borax can significantly promote the particle size of the metallic iron, which further improves the separation of the iron particles and gangue and increases the iron grade of the magnetic concentration. The final product, assaying 92.98% TFe with the recovery of 91.71%, was obtained under the optimum conditions, which can be used as the burden for steel-making instead of parts of scrap.

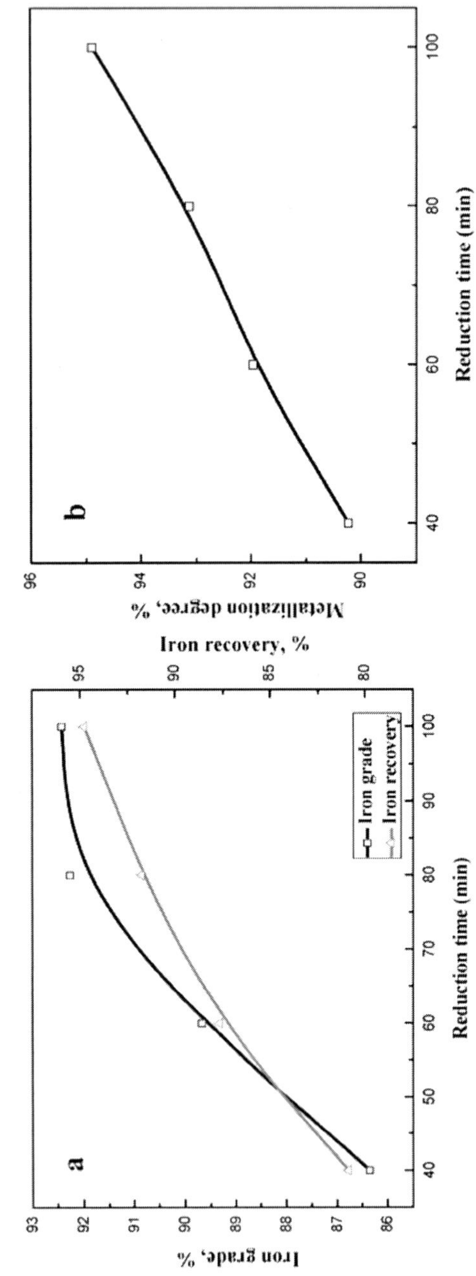


**Figure 6.** Effect of reduction time on iron grade and iron recovery of magnetic concentrate (**a**); and effect of reduction time on metallization degree of reduced briquettes (**b**) (borax dosage of 5%, reducing at 1050 °C, and grinding fineness (<0.044 mm) exceeding 68%).

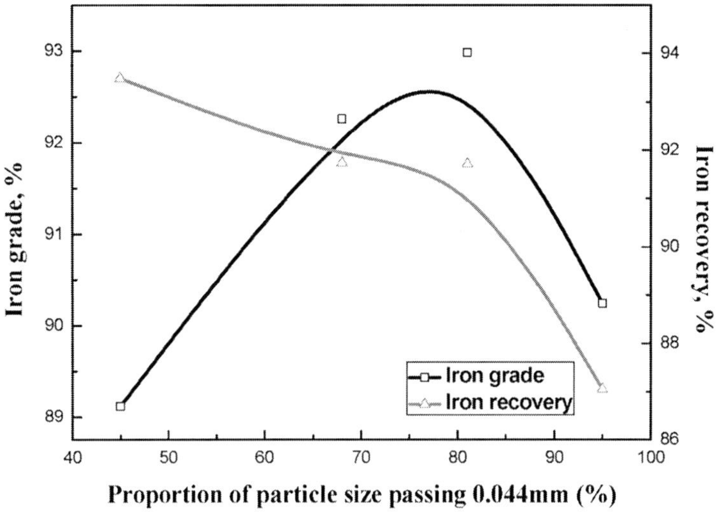


**Figure 7.** Effect of grinding fineness on iron grade and iron recovery of magnetic concentrate (borax dosage of 5%, and reducing at 1050 °C for 80 min).

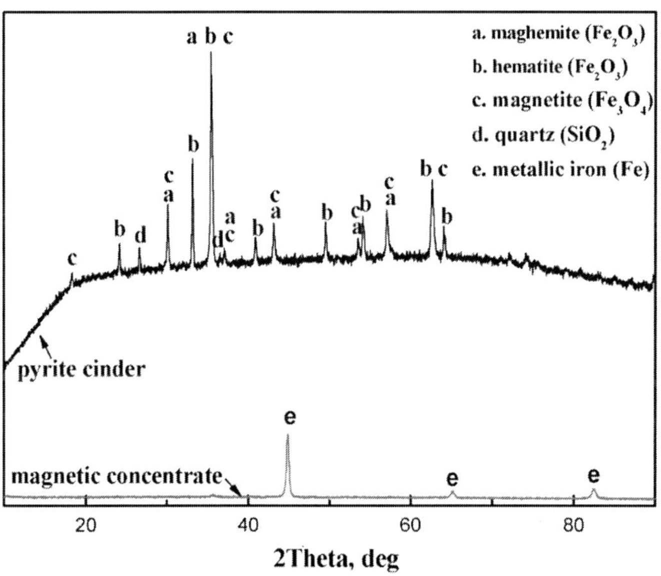


**Figure 8.** X-ray diffraction (XRD) patterns of magnetic concentrate and pyrite cinder (borax dosage of 5%, reducing at 1050 °C for 80 min, and grinding fineness (<0.044 mm) exceeding 81%).

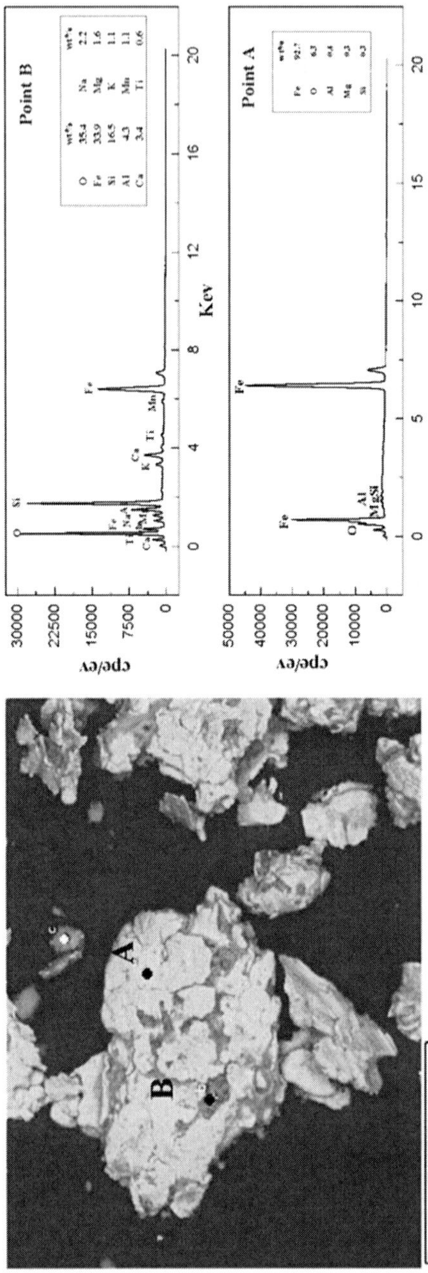


**Figure 9.** Scanning electron microscope with energy dispersion system (SEM-EDS) of the magnetic concentrate under the optimum condition (borax dosage of 5%, reducing at 1050 °C for 80 min, and grinding fineness (<0.044 mm) exceeding 81%).

## ACKNOWLEDGMENTS

The authors are thankful to the projects by National Natural Science Fund China (51504003) and Science and Technology Development Anhui Province (1501041126) and National Natural Science Fund Anhui Province (1608085QE94) and the Key Project of Natural Science Research of Anhui Universities (KJ2015A028) for sponsoring the research work.

## AUTHOR CONTRIBUTIONS

Hongming Long and Jiaxin Li conceived and designed the experiments; Qingmin Meng performed the experiments; Ping Wang and Zhanxia Di analyzed the data; Qingmin Meng contributed reagents, materials, and analysis tools; Tiejun Chun wrote the paper.

## REFERENCES

1. Zhu, D.Q.; Chun, T.J.; Pan, J.; Guo, Z.Q. Preparation of oxidised pellets using pyrite cinders as raw material. *Ironmak. Steelmak.* **2013**, *40*, 430–435.

2. Zheng, Y.; Liu, Z. Preparation of monodispersed micaceous iron oxide pigment from pyrite cinders. *Powder Technol.***2011**, *207*, 335–342.

3. Giunti, M.; Baroni, D.; Bacci, E. Hazard assessment to workers of trace metal content in pyrite cinders. *Bull. Environ. Contamin. Toxicol.* **2004**, *72*, 352–357.

4. Ye, Z.P.; He, G.W. Research on recycling technology and waste water treatment with sulfuric acid residue. *J. South China Norm. Univ.* **2010**, *2*, 72–75. (In Chinese)

5. Jin, C.; Li, D.X. Research progress on the comprehensive utilization of iron from pyrite cinder. *Metal Mine* **2011**, *10*, 162–165.

6. Bi, W.L.; Wu, W.H.; Li, J. Experimental research on recovering of iron concentrate from pyrite cinder. *Hydrometall. China* **2011**, *30*, 229–230. (In Chinese)

7. Wang, Q.L.; Zhou, H.Q.; Dai, Y.H. Technological research on desulfurizing flotation of pyrite cinder in Guangxi. *Min. Metal. Eng.* **2008**, *28*, 44–46.

8. Zheng, Z.F.; Li, M.L. Experimental study on iron recovery from pyrite cinder. *Min. Metal. Eng.* **2006**, *26*, 29–32.

9. He, B.; Tian, X.; Sun, Y.; Yang, C.; Zeng, Y.; Wang, Y.; Zhang, S.; Pi, Z. Recovery of iron oxide concentrate from high-sulfur and low-grade pyrite cinder using an innovative beneficiating process. *Hydrometallurgy* **2010**, *104*, 241–246.

10. Jin, C.; Wang, E.Q.; Li, D.X. Pressure reduction leaching iron from sulfate slag. *Nonferr. Metals* **2011**, *11*, 6–8.

11. Long, H.M.; Meng, Q.M.; Wang, P.; Chun, T.J.; Yao, Y.L. Preparation of chromium-iron metal powder from chromium slag by reduction roasting and magnetic separation. *J. IronSteel Res. Int.* **2015**, *22*, 771–776.

12. Chun, T.J.; Long, H.M.; Li, J.X. Alumina-iron separation of high alumina iron ore by carbothermic reduction and Magnetic separation. *Separat. Sci. Technol.* **2015**, *50*, 760–766.

13. Qiu, G.Z.; Jiang, T.; Xu, T.J.; Cai, R.Z. *Direct Reduction of Cold-bonded Pellets*; Centre South University Press: Changsha, China, 2001; pp. 176–180.

14. Zhu, D.Q.; Zhou, W.T.; Pan, J.; Chen, D. Improving pelletization of Brazilian hematite by adding boron-containing magnetite. *J. Cent. South Univ. Sci. Technol.* **2014**, *45*, 348–355. (In Chinese)

15. Chun, T.J.; Zhu, D.Q.; Pan, J.; He, Z. Preparation of metallic iron powder from red mud by sodium salt roasting and magnetic separation. *Can. Metal. Q.* **2014**, *53*, 183–189.

16. Chen, D.; Zhu, D.Q.; Chen, Y. Preparation of prereduced pellets by pyrite cinder containing nonferrous metals with high temperature chloridizing-reduction roasting technology. *ISIJ Int.* **2014**, *54*, 2162–2168.

17. Zhang, K.; Nieto, A.; Kleit, A. Valuation of mining operation with uncertainty and the power of waiting—A real option method. In *Mine Planning and Equipment Selection*; Springer publisher: Berlin, Germany, 2014; pp. 1503–1512.

18. Zhang, K.; Kleit, A.N. Mining rate optimization considering the stockpiling: A theoretical economics and real option model. *Resour. Policy* **2016**, *31*, 87–94.

19. Chun, T.J.; Zhu, D.Q.; Pan, J. Simultaneously roasting and magnetic separation to treat low grade siderite and hematite ores. *Miner. Process. Extr. Metal. Rev.* **2015**, *36*, 223–226.

# INDEX

**A**

active braze alloy, 180
ANOVA, 262, 263, 265

**B**

Borides, 48

**C**

Carbides, 48
carbothermic reduction, 272, 286
corrosion, 10, 139, 156, 159, 197, 199, 200, 201, 203, 204, 205, 207, 208, 209, 211, 231
Cu-Ag alloy, 180

**D**

Dental Amalgams, 2
Dental Implant Alloys, 8

**E**

electrochemistry, 180
Endodontics, 15
energy dispersive spectroscopy, 180, 185, 187

**F**

Ferrite, 146, 151, 155, 211, 227

**G**

glass substrate, 80
Green Sand Mould, 250

**H**

Hardness, 163, 166, 174

**I**

Inconel 718, 229, 230, 231, 232, 234, 235

## M

Machining, 163, 171, 177
magnetic separation, 271, 272, 274, 275, 279, 281, 286, 287
Martensite Grain Sizes, 146
metal injection molding, 229, 230, 243
Metal matrix composites, 157, 158, 175
metallic iron powder, 272, 274, 275, 281, 286
microalloyed steel, 113, 114, 115, 119, 131, 132
microgalvanic corrosion, 180, 182, 198, 200
Microstrain, 117
Morphology, 81, 98, 101, 111, 215, 222, 243

## N

Nanoindentation, 137, 155
Nickel-Titanium Instruments, 15

## P

polyethylene glycol, 229, 230, 238
polyhydroxyalkanoates, 230, 236
precipitates, 6, 18, 48, 51, 55, 57, 58, 59, 60, 61, 62, 63, 113, 164, 212, 214, 215, 219, 221
Prosthodontics, 3, 6, 20
pyrite cinder, 271, 272, 273, 274, 279, 281, 283, 285, 286

## Q

quantitative X-ray diffraction, 114, 115
Qxrd, 115

## R

Rheology, 235

## S

scanning electron microscopy, 31, 80, 114, 180
scanning kelvin probe force microscopy, 180, 204
Sigma phase embrittlement, 211
stainless steel, 9, 11, 15, 22, 210, 214, 215, 216, 217, 230, 243
Stiffness, 163

## T

Tensile Test, 37

## W

Water Debinding, 237
Wind power, 28